Renaissance II: Canadian Creativity and Innovation in the New Millennium

Edited by Richard I. Doyle

Printed in Canada on acid-free paper. ♾

ISBN 0-660-18397-8
NRC No. 44456

National Library of Canada cataloguing in publication data

Main entry under title:
Renaissance II: Canadian creativity and innovation in the new millennium

Issued also in French under title:
Renaissance II : Créativité et innovation canadiennes au cours du nouveau millénaire
Issued by the National Research Council of Canada.
ISBN 0-660-18397-8

1. Science and the arts — Congresses.
2. Technology and the arts — Congresses.
3. Creative ability — Congresses.
4. Technological innovations — Congresses.
I. Doyle, Richard I., 1952- .
II. National Research Council Canada.

X180.S3C73 2001 700.'1'05 C2001-980033-9

Contents

"To create or not to create is not a question. We Canadians will continue to create because we have done so in the past and have a tradition; but more than that, because we must work together and we must create. We have no choice."

Dr. Peter Hackett
Program Chair
Millennium Conferences on
Creativity in the Arts and Sciences

Preface

The material presented here has been collected in the context of the Millennium Conferences on Creativity in the Arts and Sciences, a series of events staged in 1999 and 2000 to help Canada mark the arrival of the new Millennium.

Some events were focused entirely on the themes of creativity, innovation, and collaboration between the arts and sciences. In other cases, individual speakers or projects were supported and presented within broader initiatives or conferences as being associated with the Millennium Conferences series.

The following pages include selected speeches as they were delivered at events such as the centrepiece conference Creativity 2000 (Appendix III), papers written to report on such events, notably the Symposium on Creativity and Innovation (Appendix I), biographical sketches, background information, quotes, and stories related to the Millennium conferences and its themes (Appendix II).

Many divergent points of view and perspectives are reflected here. However, each is a reminder of the power that lies in our imagination, our capacity to work together, our ability to learn and grow, and our humanity.

We believe that this material will help you prepare for work, play, and life in the new Millennium.

Acknowledgements

The number of individuals and organizations that contributed to the Millennium Conferences on Creativity in the Arts and Sciences is inspiring testimony to our collective commitment to creativity and innovation and a hopeful sign for the new Millennium.

Only a common desire to collaborate in new approaches toward a vision of a better future could make such an initiative possible.

From the perspective of the National Research Council of Canada (NRC), the Millennium Conferences on Creativity in the Arts and Sciences began early in 1998 with a decision to participate in Canada's Millennium celebrations through a conference series that would augment and challenge its strategies for science, technology, and innovation in the knowledge-based economy with new ideas, new partnerships, and new concepts.

Before progressing very far in its plans, NRC learned that many other national organizations were also looking to magnify their impact and contribution to society through unconventional alliances and projects.

The Humanities and Social Sciences Federation of Canada (HSSFC), for example, was considering an initiative that resulted in the special Symposium on Creativity and Innovation described in the formal Report presented in Appendix I.

Early in 1998 the National Arts Centre of Canada (NAC) was preparing a major conference that explored the role of the artist in the international human rights movement to commemorate the 50th anniversary of the signing of Universal Declaration of Human Rights, which had been drafted by Canadian John Peters Humphrey.

In the context of this expanded vision, the NAC invited NRC to collaborate in a similar conference that would bring the arts and sciences communities together in a unique forum. NRC saw this invitation as an exceptional opportunity to achieve its Millennium conferences objective.

The founding partnership of the NRC and the NAC soon grew to include the Canada Council for the Arts, the organization known primarily for its role in funding the arts in Canada, but one that has long recognized the importance of science and technology to its community.

Eventually, many of Canada's most innovative national organizations and the world's most creative minds would join the collaboration and make critical contributions to its success.

We have sought to acknowledge them all in the following pages, but know that the many projects and events touched by the Millennium conferences series could not have succeeded without the active involvement and imagination of many more individuals than have been cited here.

We are grateful to them all and particularly to those who attended these events and contributed to their success with their enthusiasm and good wishes.

Sponsors and Supporters of the Millennium Conferences on Creativity in the Arts and Sciences

National Research Council of Canada

National Arts Centre

Canada Council for the Arts

The British Council

Canadian Institutes of Health Research

BCE-Média

Canada Foundation for Innovation

Natural Sciences and Engineering Research Council

Rogers Television

Industry Canada

Social Sciences and Humanities Research Council

Humanities and Social Sciences Federation of Canada

University of Alberta

Ottawa Regional Innovation Forum

National Aboriginal Career Symposium

Canadian Aboriginal Science and Engineering Association (CASEA)

National Science Council (Taiwan)

The Embassy of Italy

The School of Dance

Canadian Medical Association

Canada Council Art Bank

NRC Steacie Institute for Molecular Sciences

NRC Institute for Information Technology

Champions for Children Foundation

University of Brescia (Italy)

Canada Science and Technology Museum Corporation

Conference on Statistics, Science and Public Policy (Herstmonceux, U.K.)

Inventive Women Project

Silicon Graphics Inc.

The Millennium Bureau of Canada

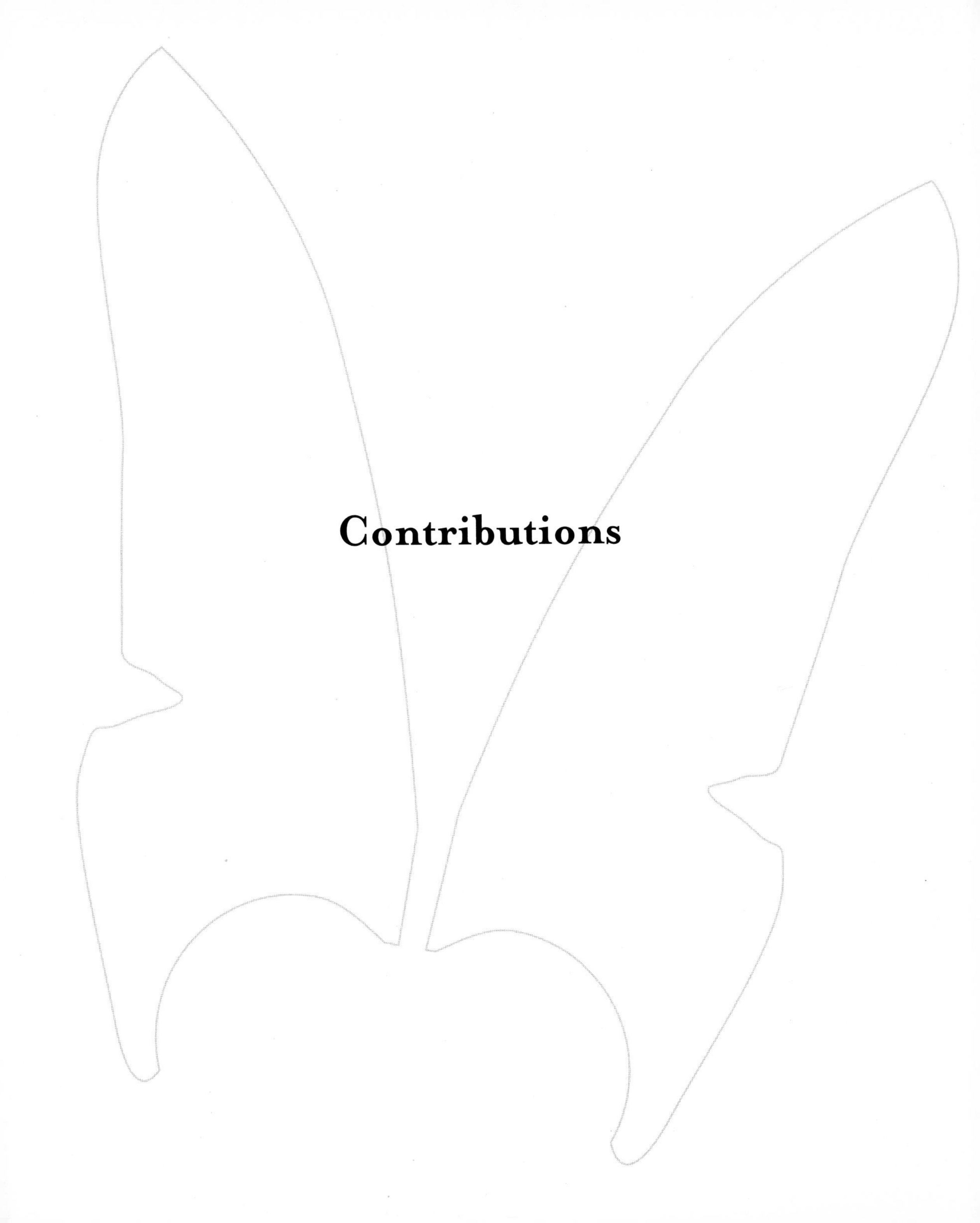

Contributions

The Governor General of Canada Calls for a New Alliance Between the Arts and Sciences

The Patron of the Millennium Conferences on Creativity in the Arts and Sciences
Her Excellency the Right Honourable Adrienne Clarkson, C.C., C.M.M., C.D.

Born in Hong Kong in 1939, Mme Clarkson came to Canada as a refugee with her family during the war in 1942. She received her early education in the Ottawa public school system and later obtained an Honours B.A. and an M.A. in English Literature from the University of Toronto. She also did post-graduate work at the Sorbonne in France.

Mme Clarkson is fluently bilingual. A leading figure in Canada's cultural life, Mme Clarkson has had a rich and distinguished career in broadcasting, journalism, the arts, and public service. From 1982 to 1987, Mme Clarkson served as the first Agent-General for Ontario in Paris, promoting Ontario's business and cultural interests in France, Italy, and Spain. She was the President and Publisher of McClelland & Stewart from 1987 to 1988. From 1965 to 1982, Mme Clarkson worked as host, writer, and producer of several influential programs on CBC Television, including Take Thirty, Adrienne at Large, and the Fifth Estate. A noted writer, she also contributed numerous articles to major newspapers and magazines in Canada and wrote three books. In 1988, she assumed responsibilities as Executive Producer, Host, and Writer for the programs Adrienne Clarkson's Summer Festival and Adrienne Clarkson Presents for a period of 10 years. She also wrote and directed several films. Her work in television has garnered her dozens of TV awards in Canada and the U.S.

Until the announcement of her appointment as Governor General, Mme Clarkson served as Chairwoman of the Board of Trustees of the Canadian Museum of Civilization in Hull, Quebec, as well as President of the Executive Board of IMZ, the international audio-visual association of music, dance, and cultural programmers, based in Vienna. She was also the Executive Producer and Host of the CBC Television program, Something Special, a Lay Bencher of the Law Society of Upper

"Perhaps it will inspire you to walk down a path which has not been lighted up by others but which you can light for yourselves. That is the meaning of all creation."

Canada as well as Honorary Patron of a number of artistic and charitable organizations.

Mme Clarkson has received numerous prestigious awards both in Canada and abroad in recognition for her outstanding contribution in a variety of endeavours. She was appointed an Officer of the Order of Canada in 1992, holds honorary doctorates from five Canadian universities, and received three honorary academic distinctions.

Mme Clarkson is married to the writer, John Ralston Saul.

In January 2000 the Governor General accepted the invitation to become Patron of the Millennium Conferences on Creativity in the Arts and Sciences. In doing so, Her Excellency lent her general endorsement to the initiative and agreed to be referenced in the capacity of Patron. The following is her address delivered at the opening of Creativity 2000 on Wednesday June 21, 2000.

Her Excellency the Right Honourable Adrienne Clarkson, C.C., C.M.M., C.D.

When I was at university some time ago, one of the streams of thought that was making the most impact on us and causing a debate was the thinking of C.P. Snow as embodied in his series of novels about life in an Oxbridge university in England. By positioning the masters of that college and also in his non-fiction writing, C.P. Snow posited the idea of the two cultures. I remember how seized we were, certainly in a debating way, with the idea that science and the arts were, to use his words, "two cultures."

He said, in his opinion, that they were separated by a "gulf of mutual incomprehension." As they lacked a "common language," scientists and literary people (which happened to be the category of the arts that he chose to focus on, being a writer himself and with a surprising lack of self-interest) were unable to speak to each other.

As I stand here with you today and among you, I can't believe that our twenty-year-old selves at the University of Toronto of that moment in time, in Canada's history, could have believed this to be true or even give it time for discussion. Here I am surrounded by artists, musicians, writers, philosophers, and scientists, and the concept of "two cultures" is hopelessly out of date.

We've been learning more and more about how our brains actually work, we've been able to bring into common parlance the studies of Jung and there now seems to be no gap between the creativity that creates science and the imagination that creates art.

Imagination, perception, intuition are all part of the way in which we now understand how we perceive and act upon the world around us.

In *The Poetics*, Aristotle wrote: *"It is not the business of the poet to tell what has happened, but what might happen and what is possible, according to probability or necessity."* We could so easily substitute the word "scientist" for "poet." And if we did that, we would understand the similarities between these areas of discovery.

For discovery is really what creativity in both the arts and science are about. We're learning now from cognitive scientists that the unconscious plays a significant role in the functioning of the brain as it creates new ideas. I am particularly struck always by the relationship of art to magic, of the act of creation as related to the totally mysterious. Perhaps this has often been downplayed in science, which is frequently perceived to be more objective as opposed to the subjective of creativity in art. But there seems to be no doubt that however we want to phrase it, the idea of inspiration or of the subconscious does play a large role in scientific discovery.

We know that the unconscious or the subconscious have a role in discovery in science. And we have examples where discoveries were preceded by moments of what we would call inspiration. Friedrich Kekulé, the 19th-century German chemist, reaching an impasse studying the structure of the benzene molecule, dozed off in front of a fire. We could ask now, "Why did he doze then? What was the meaning of that?" And then he had a vision of a snake biting its tail and in a flash he realized that the molecular structure was characterized by a ring of carbon atoms. This dream vision directed him towards his future so-called objective research. And there are so many stories akin to this.

The link between scientific discovery and creative activity has been so well established, I think, especially over the last twenty-five years, that I need not go on with any more arguments. There is simply no truth in the division between "two cultures" as C.P. Snow spoke of them to such great discussion and relative acclaim forty years ago. Purposes and processes have made the idea of supposed conflict an intellectual exercise useful to debate but not a reality to confront.

All creative people are involved in activity which is so much larger than themselves that it would be difficult for them to even state what that meant to people who have never been involved in that activity. I'm always rather amused and touched to hear both artists and scientists trying to describe to the laymen exactly what it is that they do. The familiar and dreaded question of the inexperienced interviewer "Where do you get your ideas?" is probably only second to the banality of "Tell me what you're like." What drives all creators — scientific and artistic — is passion and obsession. Most people's lives, unfortunately, contain neither or if they do, they are fixed on animal matters rather than spiritual ones.

What is very difficult for people not involved in creative activity to understand is the excitement and the awe-inspiring straining towards perfection which all discovery or creativity leads to. Einstein said that *"the most beautiful thing we can experience is the mysterious. It is the source of all true art and science."* Perhaps that is the truest thing that can be said about the activity of the people who are taking part in this Conference today.

I really want to salute the people organizing this Conference because I think they are honouring people who refuse to be pigeon-holed and refuse to stay in categories. The world loves to put people into boxes, neatly tied up. But being in a box even tied with the most beautiful ribbon and wrapped with exquisite paper is still a box. And I have been speaking here of the pure act of creativity that breaks barriers and refuses categories.

In the areas where creativity — scientific and artistic — collaborate, there are myriad examples of a synergy which leads to invention and benefit. And benefit, I may add as a footnote, is only a reasonable practical outcome in a society of fruitful collaboration.

I want to recall to you that thirty years ago scientists from around the world were exploring the idea of something called computer animation, and those who were doing so at the National Research Council had the inspiration of working with film makers at the National Film Board of Canada. By seeking the input of people who would actually use their technology, there was an extremely energizing and dramatic collaboration which in retrospect was the one that would actually give Canadians their competitive edge. We, in Canada, were the first in the world to develop computer animation. In 1997 when the scientists — and it was the scientists interestingly enough — received an Academy Award (that is the ceremony that happens the week before the one where everybody thanks their mothers!) for their technological innovation, they credited their collaboration with the film makers in the 1970s for pushing their discoveries.

Today, we really take for granted the idea that artists can use technology to create. Videos, computers, lasers are basic tools. Perhaps the degree to which scientists rely on art is less visible.

Twenty or thirty years ago, anyone strolling through the halls of the National Research Council could probably readily identify the various labs — metallurgy, biology, chemistry. Today if you did that same walk, you would be hard-pressed to discern what disciplines scientists were actually investigating. So many scientists work in the world of virtual reality, an artist-created environment that actually doesn't exist. A creation actually can bounce off the walls of technology and rebound into the world of science. Computer animation was developed for artistic purposes but now it's a valuable tool in laboratories around the world. And I haven't even begun to deal with its widespread and now essential use in computer-assisted design for such things as architecture.

Just six years ago, we designed our cottage with an architect completely through computer-aided design which not only saved us time but gave us a total idea of what it would be like to live in that limited space which we were allotting to ourselves. I thought it was marvellous then and I still do. We were given a dimensional world which we had only been taught how to approach previously by looking at Cubist paintings.

Art always shows us the way to discoveries of perception. And that is also the link between art and scientific discovery. And discovery, whether artistic or scientific, is about that mysterious passage where the voyage is everything and the end is not. I think it's wonderful that there are so many young people in the audience and they are here to listen to people who have committed themselves to their passion and their obsession. Perhaps it will inspire you to walk down a path which has not been lighted up by others but which you can light for yourselves.

That is the meaning of all creation.

Scientist and Research Manager Says We Need to Learn from Children and Look Beyond Our Place in Time

Program Chair – Creativity 2000
Vice-President (Research)
National Research Council of Canada
Dr. Peter Hackett

Dr. Peter Hackett is an internationally recognized chemical physicist who pioneered many applications of lasers in chemistry. His innovations have earned him the Noranda Lecture Award of the Canadian Society for Chemistry, the Rutherford Medal of the Royal Society of Canada, and other honours.

He pursued his post-secondary studies at the University of Southampton, England, and joined the National Research Council of Canada (NRC) as a post-doctorate fellow in 1972.

In January 1998, he was appointed NRC Vice-President of Research.

As Program Chair of the Millennium Conferences on Creativity in the Arts and Sciences, Dr. Hackett oversaw the development of the conference series and spoke formally at several events including Creativity 2000 in Ottawa and the Symposium on Creativity and Innovation in the Arts and Sciences in Edmonton.

The following is drawn from these speeches.

Dr. Peter Hackett

If the Millennium Conferences on Creativity in the Arts and Sciences produced any broad consensus or overall conclusion, it would be that creativity is without doubt Canada's most significant natural resource — in fact its primary resource.

The ability and capacity of our young people to imagine and create their futures will determine our future as a nation, including how richly we experience that future. We must value creativity and foster it in every conceivable way.

The good news is that to do so not only advances our collective interests but also values our humanity because to create is fundamental, and always has been. Just as Aristotle observed in his Metaphysics that *"All men desire to know,"* I believe that we also share a common desire to create, and this desire has driven our evolution — natural, cultural, and technological.

As discussions at the Millennium Conferences often illustrated, creativity has even made these three evolutionary types impossible to distinguish from one another. Furthermore, the rate of our technological evolution is constantly accelerating; it has now attained a pace that challenges our cultural world and our natural world's ability to cope. The integration is magnified daily.

We must therefore make a shift in our conceptual space to muster the creativity necessary to respond to the challenges before us. We must consider the whole and work together to common purpose. The encouragement of this shift was one objective of the Millennium Conferences on Creativity in the Arts and Sciences.

Rather than asking what we should teach our children, we might better ask ourselves what we should learn from them.

Creativity and Our Children

As we consider how to encourage creativity in individuals and future generations, we might rephrase the question.

Rather than asking what we should teach our children, we might better ask ourselves what we should learn from them.

Many of the Millennium Conference participants associated the creative act with childlike wonder and engagement. So before we take too much pride in reviewing the creativity of the professional adults outlined in this document, we should remember the one group that is the most creative on Earth: our children.

In this context, one of my favourite stories comes from the Hawaiian Islands in the mid-1800s, when thousands of workers from a variety of cultures were brought there to work on the great crop plantations. The adults spoke English or Japanese, Mandarin or Fijian or Gaelic, and were mutually unintelligible. But in a single generation, their children had invented and perfected a communal language, easy to learn and use but rich and expressive.

The linguist who documented this said he could no longer look at any child without a feeling of awe.

The invention of Hawaiian Pidgin illustrates that creativity uses a process that is innate to all of Earth's brainier creatures, including us — or at least to their young.

That process is play. Play is so central to the creative act that the two are indistinct, possibly identical. Like creativity, play is not merely one of our traits, even an important one. It is defining. We are who we are; we do what we do, because we play.

Both creativity and play draw upon the power of the imagination, as only the imagination can link constructive activity to unrealized enjoyment and, as in the case of Hawaiian children, to draw upon the best that each of us has to offer when we work together in common purpose.

So as we encourage our children to play and to do so creatively, we are encouraging them to use their imagination to amplify the experience and to love learning and doing.

While imaginative play does not constitute an end in itself, we might increasingly recognize it as an appropriate means of learning, exploring, contributing, and creating.

Creativity and Our Organizations

Even when adults do the inventing, their work involves creative play. Most of the epochal discoveries of science came from adults pursuing something solely for the love of it — because, as some kids would say, it was cool. The usefulness followed later.

This is important to note, as it is now generally accepted that the source of wealth, all kinds of wealth, is a creation that is specifically human: Knowledge. Knowledge applied to things we know how to do leads to productivity.

Knowledge applied to things we have not done before leads to innovation. This news has even reached the August ears of the US Federal Reserve. In June 1999, Alan Greenspan said: *"Something special has happened in the*

American economy in recent years — a remarkable run of economic growth that appears to have its roots in ongoing advances in technology."

Ours really is already a knowledge-based economy. High-tech firms now produce nearly twenty billion dollars per year in goods for export alone, employing more than four hundred thousand Canadians. In both value added and people employed, those figures exceed the most optimistic data for the entire Canadian lumber industry. The New Economy is upon us.

Decades of work by the national institutions, the companies, and the other organizations such as those participating in the Millennium conferences have contributed significantly to the national successes in creativity and innovation that built the new economy and define our quality of life.

How can we apply the notion of play and creativity in building the organizations of the future? I believe that while every great organization has its own special qualities, all are driven by two great ideas: one of creativity and another of structure. This creates a kind of tension.

Creativity flowing from bright and energetic people engaged in play is exemplified by the life of Dr. Gerhard Herzberg. The tension results from our belief that we can manage creative energy toward specific, common goals.

At the National Research Council of Canada, we see creativity as essential to our success as a research organization. It's our mission statement; it's what we do. Because of this, we have always recognized that creativity is the lifeblood of discovery, and of the innovations that flow from discovery.

Our structure, our common goal is the interest of Canada and our vision as a valued national organization. We aspire to encourage play and to manage it within the structure of a compelling vision, our collective imagination, of where we fit as an organization within Canada.

Our success as an organization lies in our capacity to set our employees free to be creative, to be their very best, and to work with others while keeping the vision of a better Canada in mind at all times.

Through the Millennium Conferences, I learned that similar combinations of creative play and tension come into play in the artistic world where the tension of discipline and technique combine with the imagination and vision to produce wonders.

Clearly, we must balance and blend these forces within our organizations and ourselves to spawn creativity.

Creativity and Our Nation

As our country and the countries of the world enter the new Millennium, they will also need to promote visions that encourage creativity, but they will need to be even more reflective, more social, more long-term than in the past.

National prosperity, though it demands all the creativity we can bring to it, is not the only challenge we face. Many other issues are as great, and we must embrace all interests and resources to realize creative solutions.

Some of our problems are local; some are global; some are both. Some regions of the world, for example, seem to be sliding into social implosion: a kind of tribalism with advanced weapons. In many regions, deforestation, drought, and other resource depletion create first poverty, then starvation, then conflict over what resources remain. At the same time, diseases escape their natural reservoirs and become pandemics that can quickly move from continental to global. In addition, technology is not a panacea. Along with its many solutions come some pressing problems.

The sheer quantity of humanity, its growing mass, is putting intolerable strain on food and energy supplies. Land, lakes, and seas are becoming cesspools. Our current state of industrialization pours poisons into the biosphere, killing whole genera and stunting normal human development. Greenhouse gases warm the Earth.

When the connected world finally has a collective intelligence, everything will change — taxes, laws and regulations, culture, knowledge, the nature of nations, and the nature of economies: everything. Yet even at the threshold of this profound and permanent revolution, it will be impossible for any one of us to have all knowledge. Indeed, it is impossible now. What should everyone know? This question demands our collective knowledge and creativity.

In the past, new solutions almost always involved new knowledge.

Tomorrow, many solutions will arise when the imagination of some creative person links existing knowledge in new ways. This will require our separate disciplines to mix more.

As nations, we must help artist to speak with scientist, scientist with laborer and poet, the businessperson with all. It is time for scientists to give full value and respect to a new perspective —

modifying the reductionism of the scientific method, and integrating the perspective of the social sciences and humanities.

The pace of change and the omnipresence of our technologies challenge us. We must respond by working on the human factors. As we do this work, we must also understand that since 'the right tool is half the job,' every creative solution is another right tool. For creativity, while often fun, is also deadly serious.

Our play is not without risk, and never has been. Nothing can be risk-free that has so much potential to change us and our world.

But it is all we can do, for 'To create or not to create' is not a question.

We Canadians will continue to create because we have done so in the past, and have a tradition; but more than that, because we must work together and we must create. We have no choice.

I believe Canada is up to the challenge because I believe that we have shown the ability to imagine and create such a future in the spirit of George Bernard Shaw's often quoted remark: *"You see things that are and ask why; I dream of things that never were and say why not."*

We create what we dream, especially when we do so together.

So if the imagination of children can encourage them to play creatively and if an organization's vision of its place within a nation can provide a context for creativity and innovation, perhaps a nation's collective imagination — its vision for itself as force among the nations of the world — can foster creativity among its citizenry.

What Canada's national vision should be will be for others to define.

Certainly, it should be inclusive, encourage us each to learn and aspire to best we can be, and foster creativity.

Sir John Maddox, the long-time Editor of *Nature*, scientist, and distinguished author, spoke at two of our Millennium Conferences events.

He suggests that the seed of such a vision lies in a focus that is longer term than many nations are accustomed to considering.

Perhaps the best way for a nation to encourage creativity within its own borders is to adopt a vision that looks beyond its own borders, its own interests, and even its own place in time.

"Current global problems all have their roots in the way in which the human race has opted out of natural selection.

For 10,000 years, we've built shelters against the weather and have industrialised agriculture so as to avoid the hazards of hunting and gathering; now, for the best of reasons, we offer sickly children paediatric medicine to avoid their untimely death.

In this artificial world, the human population grows without restraint and our emissions of carbon dioxide threaten to change the climate — and we are irrationally surprised. Some people even suppose that we can preserve our comfortable way of life and at the same time preserve all the other species on the surface of the Earth in their present condition. That's like asking that water should run uphill.

What we should be doing is to ask how long we want to keep the human race going — a century or so, a few more millennia or for the rest of time? Then we'd have a rational framework for dealing with global problems."

Sir John Maddox
Author, Scientist, former Editor of *Nature*
Quotation submitted for the Millennium Conferences on Creativity in the Arts and Sciences

Science and Engineering Research Leader Tells How Creativity Applies to Science

President
Natural Sciences and Engineering Research Council of Canada
Dr. Tom Brzustowski

Dr. Brzustowski graduated with a B.A.Sc. in Engineering Physics from the University of Toronto in 1958, and a Ph.D. in Aeronautical Engineering from Princeton in 1963.

He was a professor in the Department of Mechanical Engineering at the University of Waterloo from 1962 to 1987, teaching and carrying out research in thermodynamics and combustion. He served as Chairman of Mechanical Engineering from 1967 to 1970 and as Vice-President, Academic of the university from 1975 to 1987. After that he served as deputy minister in the Government of Ontario from 1987 to 1995, first in the Ministry of Colleges and Universities, and later in the Premier's Council.

Tom Brzustowski was appointed President of NSERC in October, 1995. Dr. Brzustowski holds honorary doctorates from the University of Guelph, Ryerson Polytechnic University, and the University of Waterloo, and has received the Engineering Alumni Medal of the University of Toronto.

The following is his address to the Creativity and the World Banquet held at the National Arts Centre (NAC) on June 20, 2000, on the eve of the Creativity 2000 conference.

Dr. Tom Brzustowski

It is my very great pleasure to offer some words of welcome at this important event, Creativity 2000, one of the Millennium Conferences on Creativity in the Arts and Sciences. There is no better way to anticipate the value of what might be achieved here than to note the names of the eminent people who have agreed to participate in the program.

This is a wonderful initiative, itself very creative, and I wish to congratulate the National Research Council, the National Arts Centre, and the Canada Council for the Arts for having organized it. NSERC is very pleased to join with them to help sponsor the activities unfolding here.

As many in the audience might know, NSERC is the Natural Sciences and Engineering Research Council of Canada. We are the agency of the Government of Canada that supports research in the natural sciences and in engineering in Canadian universities, an activity that produces what I consider to be some of Canada's most important creative intellectual work.

When I started thinking about saying a few words of welcome to this conference, I decided to find out just exactly what the word "creativity" meant and in what way it applied to science — if I was to welcome people, I needed to understand the promise of the event. It occurred to me that "creativity" might be a word like "innovation" that has a very diffuse range of popular usage, but also a narrow technical meaning that can get you into problems with some audiences.

(People can get very offended if they think they are labeled not innovative, even if they are not engaged in the activity that the economists call innovation. Might there be some similar reaction to creativity?)

I started off with a dictionary, of course (*Canadian Oxford*), and then moved on to some sources that have been valuable in the past.

What I found were these meanings for two key words:

create: I. cause to exist; make (something) new or original.......from which:

creative: I. of or involving the skilful and imaginative use of something to produce e.g.: a work of art, 2. able to create things, usu. in an imaginative way, 3. Inventive

imagination: I.a. A mental faculty forming images or concepts of external objects not present to the senses, b. the action or process of imagining or forming such images, 2. the ability of the mind to be creative or resourceful ..... which included the point of departure of one of those great circle tours through the dictionary. "Inventive" would take us on another.

These definitions underlined the intimate connection between creativity and imagination, in language that made creativity very obvious in the arts, but not so obvious in science. And yet my intuition insisted otherwise — I felt that creativity and science were strongly linked, and so I went looking for the connection between science and imagination.

I was not encouraged by what I first found in two books by an author whose scholarship and erudition I find awesome, one of my heroes, Daniel Boorstin. I thumbed through my well-worn copies of *The Discoverers* and *The Creators* and found that there were no scientists discussed in *The Creators*. Science was discussed in *The Discoverers*, and that made immediate sense, of course, since the object of basic research is precisely discovery.

My intuition started to waver.

Things didn't get any better when I turned to Daedulus, the Journal of the American Academy of Arts and Sciences.

The Winter 1998 issue, entitled *Science in Culture* carries an article by Lorraine Daston with the devastating title, "Fear and Loathing of the Imagination in Science." The paper itself is less worrisome, as the author states that *"My aim here is not to show that first-rate science requires imagination; others have already pleaded this point with vigor and eloquence. Rather, I would like to explore how and why large portions of the educated public — and many working scientists — came to think otherwise, systematically opposing imagination to science."*

According to Daston, the critical period for this view was the mid-nineteenth century. The issue seemed to be that imagination might somehow undermine the rock-solid structure of the scientific fact, that people might imagine some idiosyncratic description of nature, bend their observations to fit in with it, and proclaim them as "facts." This was not an issue of the right balance, as no positive effect of imagination on science seemed to be considered.

I find an interesting asymmetry in this concern. People were worried about imagination polluting science, but nobody seemed concerned about science polluting imagination. I have never heard the criticism that science fiction was being undermined by too much science creeping into the writing. Indeed, Jules Verne was the contemporary of some very active voices in the nineteenth-century debate about science and imagination.

But, today, at the dawn of the twenty-first century, debate about the room for imagination in science seems largely to have died down. I believe this has happened because the scientific community accepts findings as fact only through a very open, slow, and careful process, whose principal instrument is international peer review. The process involves many replications of the sequence of events: hypothesis, design and performance of an experiment, documentation of results, error analysis, improved experimental design, etc. At some point, if the sequence is clearly converging, the scientific community will declare the finding to be fact.

The critic might shout, "What about cold fusion?" but that just makes my point. The scientific community followed its processes of independent replication and peer review, and ultimately refused to accept cold fusion as fact, thus

"Creativity in scientific research is seeing what others have not seen, and making it visible to others."

demonstrating that it is perfectly capable of preventing imagination from undermining the base of scientific facts.

At this point, my intuition breathed a sigh of relief. There is no reason today to deny the link between imagination and science. But conscience nagged. Intuition still had to show that a link existed, and that it was constructive. The way to do that leapt out of the title that Daniel Boorstin gave to Part IX of *The Discoverers*: "Seeing the Invisible."

These words, I think, reveal the key to understanding the relationship between creativity in science.

I believe that creativity in science is concentrated in research, and I would hazard to define it this way:

Creativity in scientific research is seeing what others have not seen, and making it visible to others.

On reflection, that definition of creativity might work in the arts as well, if seeing is taken as the metaphor for all the senses and if it includes feeling as well. Michelangelo saw David in that great block of marble, and knew how to remove only what was hiding him from us. Monet's genius was to capture permanently on canvas for us to see forever, the instant patterns of shape, light, shade, and colour that only he could see at the time. Writers of fiction make the constructs of their imagination visible to us in the written word. Composers turn the sounds they imagine into written music that performers can read, and then conductors and players transform notes that most of us can't read into sounds that move us.

In science, seeing in the metaphorical sense is very important.

Scientific research reveals new relationships — correlations to start with, then eventually causal links as well. Science provides new descriptions of nature that we must see with our minds more than our eyes: Einstein's curved four-dimensional space–time to portray the action of gravity is an obvious example. For another, the chemist Kekulé dreamt of a serpent eating its tail, and made us see the invisibly small benzene molecule as a ring structure. Many famous conjectures in mathematics, that ultimately turned out to be enormously important, were accessible statements of results that were imagined long before they could be proved.

But seeing in the physical sense is, and always has been, a very important outcome of scientific research. Three centuries ago science gave us the telescope and the microscope.

Over the last three decades, it gave us the CAT scanner and the MRI machine, so important for seeing within ourselves; views of the outer reaches of the universe — and thus some of its earliest moments — showing not only the images of distant galaxies but also their velocity away from us; atomic force microscopes that reveal atoms lined up in crystals; multi-hued visualizations of the pressure exerted by air flowing past a high-speed aircraft that is easily modified because it exists only as a mathematical construct in a computer; and many other images that are often as striking in their beauty as they are in importance.

So there, that's settled. I have managed to reassure myself that imagination and creativity are very important in science. That justifies my words of welcome.

Now, I can eat my dinner with a clear conscience, and later sit back with my intuition intact and learn from another of my heroes, John Maddox.

Thank you.

National Arts Centre Head Says Artistic Capacity Will Impact on Knowledge-Based Economy

Director General and CEO
National Arts Centre Canada (NAC)
Peter Herrndorf

Peter Herrndorf has had an illustrious career in broadcasting and the arts in Canada.

He is the former Chairman and CEO of TVOntario, and served as Vice President and General Manager of CBC's English Radio and Television networks in the early 1980's.

He is also the Founding President of the Governor General's Performing Arts Awards Foundation; the past Chairman of the Board of the Canadian Museum of Civilization; the past Chairman of the Board of The Canadian Stage Company in Toronto; the past Chairman of the Stratford Shakespearean Festival Board of Governors; and the founding President of the Toronto Arts Awards Foundation. In the fall of 1995, the Prime Minister appointed him to a three-member Mandate Review to review the mandates of the National Film Board, Telefilm, and the Canadian Broadcasting Corporation.

A recipient of numerous awards and honours, Peter Herrndorf received an honorary Doctor of Laws degree from York University in 1989 and from the University of Winnipeg in 1993. In 1998, Mr. Herrndorf was awarded the John Drainie Award by the Academy of Canadian Cinema and Television for his distinguished lifetime contribution to broadcasting. In 1999, he was chosen as a lifetime "fellow" of the Ontario Teachers Federation for his contribution to the educational life of the Province of Ontario while he was at TVOntario, and was named as a "Distinguished Educator for 1999" by the Ontario Institute for Studies in Education (OISE).

Appointed in 1993 as an Officer of the Order of Canada, Peter Herrndorf was the first recipient of the William Kilbourn Award presented by the Toronto Arts Awards Foundation for lifetime contribution to the arts in Toronto in 1995. In June 2000,

Mr. Herrndorf was awarded the Diplôme d'honneur by the Canadian Conference of the Arts (CCA).

An active supporter of the Millennium Conferences on Creativity in the Arts and Sciences, Mr. Herrndorf spoke in support of new alliances between the arts and sciences communities on many occasions in 1999 and 2000. The following are his remarks at Creativity 2000, which include an opening address and an introduction to the afternoon session entitled "Reconfigurations: Structure and Space" as well as comments made by some of the participants in these sessions.

Peter Herrndorf

Opening Remarks

We're very proud to work with the National Research Council and the Canada Council for the Arts to bring together so many people with a passion for the arts and sciences. We're proud that the NAC has been chosen as the venue

Why here?

Why not gather across the canal at the Congress Centre to discuss creativity? Or across the street at the Chateau Laurier ballroom?

This afternoon, I'm going to have an opportunity to talk about the science of why this is such a great place for us to gather.

But right now, let me get mystical about why we're here.

On this stage, the finest dancers of our times have performed. As you listen to the sound of my voice and my words, think about other voices and other words heard in this theatre — the greatest actors expressing the most subtle and profound thoughts from the greatest dramas of not just our time, not just the past hundred years, and not just the past millennium — the finest theatre going right back to the Ancient Greeks. If a space can be infused with spirit — if there are « vibes » that linger when the physical presence is gone — then this is a very powerful place. Combine it with the Studio Theatre next door, and the Southam Hall at the other end of the building.

In the 30 years since it opened, these spaces have had perhaps the highest ratio of talent-per-square-inch of any space in all of Canada.

And now that you're all here, the ratio has just gone up!

The National Arts Centre believes passionately in what we have come to discuss today. We believe there is a community of interest among everyone who enjoys the adventure, the exploration, and the fun of creativity — whether your means of creative expression is in the sciences or the arts.

We believe the arts and sciences have much in common and we look forward to examining what we have in common today.

And we believe that Canada's future, in a knowledge-based economy, will be shaped by creativity and innovation. The arts stimulate the brain's synapses.

They get us thinking about things in new ways. The artistic capacity of a nation will have a direct impact on its ability to compete in this knowledge-based economy. For that reason alone — and there are many other reasons as well, I assure — but for that reason alone, Canada's public policy makers must champion the arts and sciences.

It's my pleasure, now, to introduce one such champion.

The Honourable Herb Gray has served in Parliament longer than any sitting member.

He has had a stellar political career since he was first elected to the House of Commons in 1962, and serves as the Member for Windsor West. He has served in ten Cabinet portfolios, as well as Leader of the Official Opposition, and is now Deputy Prime Minister. Prime Minister Chrétien also gave him responsibilities for the Millennium Bureau of Canada. Mr. Gray has come to the NAC many times to enjoy the performances here. Minister, I know that your tastes in the arts are very eclectic. I know that you're a great fan, for example, of Bruce Springsteen, who performed at the NAC back in the 70s.

You could probably tell this audience, even better than I could, about the range of talent that has graced this stage. It's an honour to have you bring us greetings from the Government of Canada.

> *"It is the federal government's view that not only must we make wise investments across the entire spectrum of disciplines in fields of knowledge, in both the physical and social sciences, but just as critical is the creation and maintenance of the links between them."*
>
> **Honourable Herb Gray**, P.C., M.P.
> Deputy Prime Minister of Canada

Reconfigurations: Structure and Space

Welcome back from the break. We've enjoyed a very stimulating discussion this afternoon about how the brain creates.

"The arts stimulate the brain's synapses. They get us thinking about things in new ways."

Now we move into a discussion about structure and space.

Let's start by listening to space: listening to how structure affects sound. I want you to stop listening to what I'm saying for a moment, and listen to the sound of my voice. Listen to how it reverberates in this theatre. Listen to how it fades.

I am demonstrating theatre acoustics — a field that fascinates architects and actors, physicists and sound technicians. A room is a musical instrument. It resonates with the sound within it. It adds quality and tone to the sound.

People interested in acoustics speak in terms of diffusion, intimacy, spaciousness, and bass ratio — qualities that affect what kind of art can be performed in a particular space. Music has evolved in response to the sound qualities of the space in which it is performed.

Gregorian chant was performed in stone cathedrals with soaring ceilings, where sounds take five to ten seconds to fade away. The music suitable for that kind of space is slow and dreamy, full of open vowels.

Hundreds of years later, music was written to be performed in small rooms, with plaster walls, where it took sound only one-and-a-half seconds to fade. Composers like Bach and Vivaldi wrote music full of intricacies and polyphany. It sounds great in small rooms. It doesn't work in cathedrals.

Do composers know the science of sound when they write? Not likely. But they know intuitively what kind of music works in the rooms in which it will be performed. Did the builders of medieval cathedrals or baroque palaces think about how their buildings affected what could be performed there? Not likely.

But today we do know how the science of space and acoustics affects the art of creating music. Scientists, technicians, and architects still don't have all the answers to what makes great sound in a room, but they're getting closer. This is a 20th-century example of how scientists, engineers, and artists have been collaborating.

But what about the 21st century? Technology is creating entirely new challenges and opportunities for artists and performers, scientists and technicians. In London, Ontario, the NRC has a Virtual Environment Technology Centre. Step inside its theatre, and you may find yourself in the prototype of the car of the future, or the inside of the human heart, or the streets of Renaissance Florence. Now that's a theatre!

Today, the centre is used by designers and engineers. The technology is available to the entertainment industry. What kind of art will evolve to take advantage of the ability to create an entire new reality for anyone who steps inside a theatre? What skills will artists need to master? What features will the designers of the technology need to include? I don't have the answers. We need a dialogue between arts and technology about what is possible; between producers and audience about what is desired.

That kind of discussion constitutes a small part of the topic we will look at in Module 3 of our Creativity Forum.

How do the elements of structure and space influence our thinking? What can human thought do to manipulate structure and space? I'm pleased to introduce a panel of very distinguished individuals who will approach the topic from very different directions.

Sir Harold Kroto won the Nobel Prize for Chemistry in 1996.

His research involved a new class of carbon molecules, C60, known as fullerenes. Earlier in his life, however, he was pulled very strongly toward working with another type of carbon altogether: graphite, also known as pencil lead. He seriously considered a career as a graphic artist. Audiences in his native Britain are very familiar with Sir Harold as a science communicator and film producer whose programs are featured on the BBC.

Sir Harry has strong connections to Ottawa and the National Research Council, where he held a post-doctoral in the mid-60s.

"If you have faith... if you actually believe that you are going to get there, you are going to get there."

"Whatever things you like, three, four, or five things, just do each of those to the best of your ability, and don't give up."

"Even though you may not think that you are particularly good at it, if you really do it as best you can, you will probably be better than someone who could have done it better, but didn't."

Sir Harry Kroto
Nobel Laureate in Chemistry
Creativity 2000

The work of our second panelist is very familiar to Ottawa. I have had the privilege of working with our next panellist when I was Chairman of the Canadian Museum of Civilization.

Douglas J. Cardinal is the architect responsible for that museum, as well as many other outstanding buildings in Canada and around the world. In 1993, his firm was awarded the design commission for the National Museum of the American Indian. His firm pioneered the use of computers in architecture, and remains at the forefront of applying technology.

"We all have this marvelous gift of creativity; we all are magical beings. The problem is that we keep ourselves small by not taking responsibility for the individual power that we all have as creative beings. We operate from fear rather than commitment."

"When you are able to make a commitment and say: 'we are going to do this no matter what,' and you become completely unreasonable, then your visions occur. I almost have to discard reason to make these things happen."

Douglas Cardinal
Architect
Creativity 2000

To facilitate the discussion, we have invited **Don McKeller**. I have admired Don's work as an actor, writer, and film maker for many years. Most recently, Canadian television audiences have seen him in his second season of *Twitch City* which he also created and co-wrote.

The film community knows his directing work through the feature film, *Last Night*, which won the Prix de la Jeunesse at Cannes in 1998. I urge anyone who enjoys the music of the NAC Orchestra or our Baroque series to see another film in which Mr. McKeller starred and shares writing credits, *The Red Violin*. I now turn things over to Mr. McKeller.

"When it comes to creativity, I don't think that you should underplay resistance to structure. Structures aren't always bad; if I didn't have anything to react against, I wouldn't be able to do my work."

Don McKeller
Film maker
Creativity 2000

Innovative Architect Says We Must Confront Our Fears as Enlightened "Spiritual Warriors"

Architect
Douglas Cardinal

Douglas J. Cardinal began practising architecture in Edmonton in 1964. In January of 1976 he incorporated under the name of Douglas J. Cardinal Architect Limited. Since its inception, the firm has undertaken projects of a diverse nature — from individual houses to institutional and governmental projects to twenty-five year community development plans.

Douglas Cardinal is known nationally and internationally for his signature architectural designs, demonstrated in projects such as his award-winning St. Mary's Church in Red Deer, the Grande Prairie Regional College, the Ponoka Provincial Building, St. Albert Place, the Edmonton Space Sciences Centre, and the Canadian Museum of Civilization, a facility to display and house our national treasures. The firm's work also includes the Kahnawake Tourist Village for the Kahnawake Mohawks in Montreal; the Saskatchewan Indian Federated College in Regina, Saskatchewan; a major hotel complex and Children and Elders' Center for the Oneida Indian Nation of New York near Syracuse, New York; and a master plan for the Cree village of Oujé-Bougoumou, Quebec. The village has earned a United Nation's Award for sustainable development. In 1993 the firm was awarded the design commission for the National Museum of the American Indian on the last remaining site on the Mall in Washington, DC. The firm was a pioneer and world leader in the use of computers in the profession and the business of architecture, and has constantly kept up with technological advancements in this field. We now operate a fully integrated office-wide network, employing the latest CAD technologies in all our work.

Douglas Cardinal is a recipient of the Order of Canada, Canada's highest honour. In 1999 he was awarded the Royal Architectural Institute of Canada Gold Medal, the highest award the profession of architecture in Canada bestows. The firm currently has its head office in Ottawa, where it moved in 1985 to complete the Canadian Museum of Civilization. Douglas Cardinal spoke at several events associated with the

Millennium Conference on Creativity in the Arts and Sciences including the Symposium on Creativity and Innovation in Edmonton, Alberta, and Creativity 2000 in Ottawa. The following are notes he prepared for his presentation at the latter.

Douglas J. Cardinal

We are magical beings, as human beings, because we have this power of creativity. No other beings on this planet have this gift because their patterns of existence are set and just as a deer is always a deer, a bear can only be a bear, a fish only a fish. But, we can recreate ourselves and create the tools to be anything we want, faster than a cheetah, fly higher than a bird, stronger than an elephant; we're not limited. All we have to do is declare our intentions powerfully and keep our word, not operating on reason, but on a total commitment and we make our visions happen. For example, the Wright brothers: The United States government spent $600,000.00 to scientifically prove that machines heavier than air could not possibly fly. However, the Wright brothers, who operated a bicycle shop, had an absolute commitment to fly, and believed that flying was possible. They invented the first flying machines. They weren't scientists, but they were creative people.

Creativity is creating something from nothing. So, entering the creative domain you must be willing to stand on what is known and leap into the vast abyss of the unknown. The known universe has already been created. The creative world is the vast abyss beyond what is known. It is a land of total possibility, a blank sheet of paper that is the land of the eagle, where our true power as a human being "dwells," is found, and is expressed.

The knowledge that we have that is defined I believe is too limited to solve the problems we face today. I believe we must go beyond the created knowledge we have today and be willing to leap into the unknown, to a land where all possibilities can occur.

Einstein developing his theory of relativity made this leap of recreating a whole new way of looking at the universe, although the world around him lived in a universe created by Newton. All of these creative people that have created the vast knowledge we have today took the personal responsibility of expressing this marvelous gift that each one of us has. The reason that most of us do not exercise or use this powerful gift is that we operate from fear. We are terrified of looking bad, of failing, of being ostracized from the group. Fear is unlimited. It is our fear that keeps us small.

We operate from fear and we are so fearful we do not even take responsibility for our lives, much less the powerful gift we have. We give responsibility away to others, to all the institutions that we have created, to people that want to take the power and responsibility from each of us to make vast institutions that control our very lives. All of this is based on our own fear of taking responsibility. Not only do we foster fear within ourselves, we also support all the people and all the institutions around us that magnify the fears within us. Where would the Church and the government be? All of these institutions are based on our not taking responsibility for our lives. All of these groups and individuals that we give our power to have a vested interest in keeping us small.

Ultimately they do not serve the public. These institutions only serve to serve themselves. If we took full responsibility for ourselves, for people we touch with our lives around us, and make a contribution that we're all capable of, we would truly serve our families, our communities, our governments, all the institutions of our society, and ourselves.

Why do we run our lives with fear? Why do we give up responsibility? For the individual power we each possess? Why do we have such a lack of trust in others and in ourselves? I believe it is tied to the fact that we are mortal beings. We're terrified of our own death, of our own demise. There is a sense of helplessness and anger that we are terminal and want to survive at all costs, and in order to function in our lives we deny our own mortality. We have not come to terms with our own death. I believe when you come to terms with your own death and realize you only die once, that it is inevitable, and realize that you are not going to get off the planet alive anyway, so why not concentrate on living to the ultimate and use the precious time that you have. This life is a gift to make a contribution to your own growth and development and those around you. Why be a coward and die a thousand times each time you sacrifice your potential out of fear.

When you walk with death it can become your friend because it will remind you that you have not time to mess about. If you take on your worst enemy, your fearful self, and trust the marvelous gift and power that you have within you then you can walk in trust with yourself and in trust with everyone around you. Thus you have less fear of death than fear of wasting your life, wasting this precious gift of life. Then you will be a fearless warrior, and as a fearless warrior you will gain individual power.

However, to be a knowledgeable being, responsible for your own evolution and that of others around you, you can be limited in your contribution by misusing your power. Power is corruptible and it takes even more warriorship to deal with power than to deal with fear. If you are willing to take on that battle with yourself with a true warrior's stance then you will realize that a

"... we are so fearful we do not even take responsibility for our lives, much less the powerful gift we have ... creativity."

truly powerful person does not need the power of others. Indeed he has the opportunity of empowering others to make a contribution in their lives. Power is something to give away. You do not seek power, instead you seek enlightenment. Power will find you to empower yourself and others to truly make a difference in your lives, and to always walk in trust serving others.

People that seek power and need the power of others are not whole human beings. When you give power to people like Napoleon, Hitler, Mussolini, Franco, or the many tyrants we have found in our society, we create the misuse of power, which tyrannizes every one of us. Just to be a balanced person, Napoleon was an incomplete man who needed all the power of others and of a horse that was 17 hands tall. Anyone needing the power of others is an unbalanced person, an incomplete human being. They are the last people to whom we should be giving our power. Never give away your responsibility and power to an individual or a group.

They can only misuse it. Use your power to empower others in their own lives. But never give your power and responsibility away. You need everything you can to meet your own commitment and bring forth your own contribution, which is your responsibility. You need all the power that you can to defeat the voices within you that can keep you small.

To silence voices so that you can be a luminous body, a strong creative life force that can make a difference. When you regard yourself as more than physical beings, see your self as a powerful light housed in a physical container, then you can truly release your creative energy and bring humanity one step forward in their evolution.

I remember Buckminster Fuller telling us all that his contribution, his creativity only became alive when he saw himself as the water in a throw-away plastic cup; that he was not the plastic cup, rather he was what was in that cup. As soon as he saw his physical body as a throwaway he was able to use that powerful light that was himself to fulfill his creative endeavours.

In my culture freedom has always been paramount. We did not want to be part of agrarian culture although it was far more comfortable to be part of an agrarian society. These agrarian societies meant that you had to give power to an elite group that became tyrants terrorizing each one of its members. This created monolithic structures dictated by a few who equated themselves even to the power of the sun or to an unearthly power.

In the plains we preferred to be hunters and gatherers, which to survive was a much greater challenge. To survive we had to learn to be at one with the earth and all of nature, particularly at one with the animals that gave us life. We had to be at one with the deer, the buffalo, to know their habits. We had to be like them to survive. We could not be above the earth or other creatures.

We had to be at par with a blade of grass. When you have become at par with a blade of grass you can start communicating with all of the living beings. When you have become a deer and know all his patterns he can easily fall into your traps, as though he had offered himself to you. There was reverence — for the animal that gave his life so that you can have yours. This is what you learned when you became a good hunter, a good provider.

When you became trained as a warrior you realized that you had to gain further knowledge of your fellow man. You had to be patternless to be any creature you wished to be. In order to be either a deer or buffalo you change patterns. If you did not change and become patternless you could fall into the traps of your fellow human beings. You had to live outside the realm of your predictable patterned human behaviour. A good warrior was like a shaman who could change himself at will. His battle was to fight tyranny. The tyranny in himself and the tyranny in others, particularly those he had to defend against who may jeopardize his family or his hunting grounds. Indeed, to evolve to be a more knowledgeable person he was to seek tyrants to do battle, to learn and evolve. He regarded the ultimate tyrant as the Great Spirit who could take his life and spirit. It was only the petty tyrant who could take life and physical being because petty tyrants only release the spirit. The small petty tyrants of today can only take material goods. Then the little tiny tyrants make life miserable. In all instances your stance to take on tyranny is to never back down, willing to give life when you make a stand.

Then there are the enlightened spiritual warriors.

This is an even greater challenge because spiritual warriors take on the tyranny within one's self. A spiritual warrior must go through ceremonies to taste death and come back from death. In doing so he learns that it is a tragedy to take the life of another human being for any reason. When you become at one with all living beings you realize that the only creatures on the plains that take the life of their own species or territory is the human animal. No other animal on the plains takes the life of its own kind. Wolves bare their throats to their victors, the victors do no kill. Battle between animals on the plains is a sign of strength. Death is usually accidental.

To be a knowledgeable human being walking with his creatures is to be a spiritual warrior. On the plains, spiritual warriors rode to battle with coup sticks with which they would touch a vulnerable part of their adversaries to say: "I can kill you, but I spare you, you are a son, a husband, a father. I will show you my strength so that you will gather in a circle with a sacred pipe. We will agree on our territory, knowing our individual power and strength to protect families, in a peaceful way like other creatures."

Spiritual warriors rode into battle as human beings not human cannibals.

It was putting one's life on the line to trust one's self, to follow the path of enlightenment. But most importantly it was putting your life on the line to completely trust your enemy with your life. Believing that if you projected yourself as a human being that you would create that human being with your enemy, which is the ultimate trust of the goodness in your fellow man. If any person of the tribe killed a spiritual warrior, they were a shameful person.

I believe this traditional path of a man of knowledge who seeks enlightenment is a way of knowledge that is even more important in the world today. It is time to share our knowledge, our traditions and our values with our brothers and sisters, all the members of the human family in this global village.

It may be against our values and beliefs to be part of an agrarian culture, or part of the industrial age that has caused so much devastation to the planet, but now is the opportunity to embrace the information age with the knowledge that we have survived in this land for thousands of years.

Rock Musician and Scientist Says the Truly Creative Are Visual Thinkers Who Begin by Imagining

Information Technology Researcher, Composer, Musician, and Arts Leader Paul Hoffert

Paul Hoffert, a scientist and information technology enthusiast as well as a composer and musician, was a young researcher working in the National Research Council of Canada (NRC) laboratories in Ottawa when his groundbreaking rock band, Lighthouse, was conceived in the late 1960s.

His NRC colleagues at the time included the inventor of the music synthesizer Hugh LeCaine and computer music pioneers such as Ken Pulfer. He had already established a reputation as a music technology innovator at NRC when he was approached by drummer Skip Prokop with the concept of a rock band that included string and jazz-style horn sections. In the early 1970s Lighthouse was the top rock band in Canada winning the Juno award for Group of the Year four years in a row, racking up nine gold and platinum records, and selling out concert halls around the world.

Hoffert is exceptional in the diversity of his interests. His resume includes leadership positions in both the arts and sciences: Executive Director of CulTech Research Centre and a champion of the wired world and new information and telecommunication technologies, Adjunct Professor at York University, Research Professor at Sheridan College, and former Executive Director of Intercom Ontario, the consortium behind the first totally wired, interactive community in the world.

A former Chair of the Ontario Arts Council, Founder of the Canadian Independent Recording Producers Association, President of the Academy of Canadian Cinema and Television, Director of Canada's Performing Rights Organization, and Executive Producer of the Gemini Awards TV broadcasts, Hoffert is clearly a leader in the arts world as well. He has received many honours for his music including the San Francisco Film Festival, Genie, Gemini, Clio, SOCAN Film/TV Composer, and Juno awards both as a member of Lighthouse and for his solo classical music work, the Hoffert Violin

"The truly creative and innovative are not stymied by coming to the end of a road beyond which there is no clear path or destination. They imagine the goal they want to achieve, set it as a destination, and then figure out how to build the appropriate road to reach it. They are not bound by deductive reasoning ..."

Concerto. He is a Member of Canadian Rock & Roll Hall of Fame.

Paul Hoffert gave the keynote address at the Toward a Century of Creativity and Innovation Awards Banquet at the Canadian Museum of Civilization in Hull, Quebec, in February 2000 as part of the Millennium Conferences on Creativity in the Arts and Sciences.

Paul Hoffert

I often have the great pleasure of speaking at a wide variety of engagements, but there is something very special about this one, as it brings me back to my old stomping grounds.

I originally got involved with this great institution by chance. After my studies as a triple major in math, physics, and chemistry at University of Toronto, I followed my muse and formed a rock and roll band. In the late 60's, we performed frequently in Ottawa, and I used to drive by the buildings on Montreal Road, wondering what sort of research took place there. One day, I stopped, walked into the lobby and asked the first person I saw, *"What do you people do here?"*

The retort was, understandably, *"What exactly are you doing here?"* I explained my double background in the arts and sciences and my curiosity about how I might blend those in a research setting. To my surprise and joy, they didn't kick me out. Instead, I was shown a series of labs and introduced to researchers that were devoted to computer graphics, computer music, and acoustics. Within a few months, I had a username and password on one of the NRC computer systems.

During my time here, I had the honour of working with some of the many legendary scientists who have come to the NRC. Bill Buxton, now director of research for Alias Wavefront, was one of my NRC colleagues in those days, as was Ken Pulfer, computer music pioneer and longtime NRC scientist.

They told me about Hugh LeCaine, an eccentric genius at NRC, who spoke fluent Swahili and invented new music machines. Hugh was a fascinating guy who invented the first music synthesizer (that's right, at the National Research Council). He named it after the ancient musical instrument called the Sackbutt because he had an odd sense of humor and knew that only a few musical literati would make the connection and get the joke.

The original Sackbutt did sound a bit like a medieval instrument and made a rasping sound. It looked, however, like a remnant of the industrial revolution. But it combined electronic processing with tape recordings of real sounds that you could manipulate in a musical manner. Today, rap artists like Puff Daddy use sampling synthesizers in every one of their hit records — descendents of Hugh's original work. Hugh's innovations in the Sackbutt and in other variable-speed multi-track recorders and touch-sensitive keyboards embodied the principles of the NRC, just like the inventions of more orthodox NRC researchers. He was absolutely brilliant and generous with his ideas.

Now the NRC tolerated Hugh tinkering with music because they understood that the arts and sciences have much that connect them. The project I was working on involved music, graphics, and input devices for computers, including piano-like keyboards and a mechanical mouse that predated Apple's mouse by a dozen years. As many great scientists, such as Einstein, have noted, that "Eureka Moment" in which an image is formed in the mind that solves a thorny problem, is shared by great artists and scientists alike.

A painter envisions a finished canvas, a composer "hears" a song, and Einstein imagined time as a fourth dimension. These right-brain exercises illustrate how innovators in the arts and sciences share more with each other than they do with the left-brain focused engineers who bring their imaginations to commercial success.

The NRC is one of the few organizations that tolerates this linkage of creators — I want to encourage them, and you, to continue to think of the arts and artists as partners in the future that we're inventing together.

Here's a concrete example of how musicians benefited from research at the NRC. When I formed my band in 1969, we faced an acoustics problem that was a formidable obstacle to reaching our audiences. Lighthouse was a rock band with drums, guitars and amps, but also incorporated a jazz horn section and a quartet of strings.

On stage, the violins, viola, and cello could never be properly mixed above the din of the traditional rock instruments. We had many consultations with PA professionals at the time who were unable to offer viable solutions for amplifying our strings without getting the inevitable acoustic feedback.

After bringing some of my string players to NRC and discussing acoustic and electronic issues, they began collaborating with instrument designers who came up with new solid-body stringed instruments whose output was routed to digital signal processors to simulate the sound of natural stringed instruments.

These sounded so good that a spin-off company was formed by Dick Armin, our cellist, who continues to supply these high-quality electronic instruments to some of the world's best string players, including cellist Yo-Yo Ma. The tangible result for us? Our fiddles rocked as loud as the drums!

Earlier, Dr. Carty touched on the concept of genius. Here's what Charles Baudelaire, the French poet from the 1800s, said about genius.

"Genius is no more than childhood recaptured at will, childhood equipped now with man's physical means to express itself, and with the analytical mind that enables it to bring order into the sum of experience..."

Baudelaire is speaking of the genius' openness and naiveté, like those of a child, playing with toys to help illuminate more complex processes. For a genius, creativity isn't a hobby. It isn't something you squeeze between rush hour drives and chauffeuring the kids to hockey practice. The great ones spend countless hours honing their craft, learning to be better scientists, better musicians, and better hockey players.

When we admire genius, we do more than admire talent. Talent is what you are born with — the deck of genetic cards dealt to you by the chance arrangement of your parents' DNA. There's nothing admirable about that. What inspires us is not raw talent, but the application of that talent by dedicated development and nurturing.

Mark Twain once said, *"Thousands of geniuses live and die undiscovered — either by themselves or by others."*

That's especially true in a modest country like ours. Canadians are famous for not wanting to seem too big for their boots.

How many Canadians, for example, know about the century of innovation we have seen in Canada's federal government labs?

How many know about the seminal roles Canadians played in developments ranging from insulin to frozen food? How about the TB vaccine, or the pacemaker, or the Geiger counter, or the flight recorder?

The list goes on — and I've only been talking about federal labs. When you include corporate and university labs, Canadian scientific accomplishments fill the horizon and beyond.

We've all seen those Heritage Minutes honouring Canada's history. If these were dedicated to science, we could fill many Heritage Hours not minutes.

This evening, I am especially honoured and pleased to be part of this banquet under the auspices of the Millennium Conferences on Creativity in the Arts and Sciences. In this respect, I would like to commend the National Research Council, the National Arts Centre, the Canada Council for Arts, and their other partners for pointing out how artificial the division between the arts and sciences communities really is and for encouraging a new, Canadian-led renaissance.

I have believed for many years that the compartmentalization and specialization introduced by the Industrial Revolution has in many ways hampered creativity and innovation. In fact, creative and innovative people in both the arts and science communities have much in common.

I know I touched on this earlier, but for me, this is a fundamental issue that has been pushed aside by the forces of specialization. The truly creative and innovative are not stymied by coming to the end of a road beyond which there is no clear path or destination. They imagine the goal they want to achieve, set it as a destination, and then figure out how to build the appropriate road to reach it. They are not bound by deductive reasoning, by summing all the known facts.

I am both an artist and a scientist. I know that art touches people in a special emotional way, while sciences stimulates us intellectually. But when the two elements come together, they create something truly wonderful. They make magic.

And this is part of my memory of NRC, which was known as a unique place in North America, if not the world, for respecting the creative and innovative power of arts and sciences. We Canadians have led the world before — through synthetic music, computer animation, and acoustical sciences. There is no reason we can't do it again. There is no reason that Canada cannot be the cradle of a new renaissance that reunites the gifts of artists and scientists.

Happily, we are beginning to take steps to recognize the creative genius in Canada's labs. And tonight, we're recognizing the creative genius at the NRC.

You'll remember Mark Twain saying that sometimes geniuses don't recognize themselves as geniuses. Tonight we will change that. Tonight we are saying, *"Yes, you are important to us. Yes, your work has made a difference. Yes, your work has been noticed."*

We are emerging from one of the most exciting centuries that art and science have ever seen. And as we'll see tonight, the best is still to come.

Tonight, we honour you, the engine of our country's future. What you accomplish in your labs and workplaces is essential to the fabric of the nation.

Arts Leader Talks of the Link Between the Brain and the Creative Mind

Director
Canada Council for the Arts
Shirley L. Thomson

Dr. Thomson is a leading figure in the Canadian arts community.

From 1987 to 1997, she served as Director of the National Gallery of Canada, during which time she oversaw the Gallery's move into its new home, an outstanding series of exhibitions and acquisitions, and its change to Crown Corporation status.

From 1985 to 1987, Dr. Thomson was Secretary-General of the Canadian Commission for UNESCO in which position she also sat on the Senior Management Committee of the Canada Council. From 1982 to 1985, she was Director of the McCord Museum in Montreal. Dr. Thomson has received several honorary doctorates. She was named a "Chevalier des arts et des lettres" by the Government of France, and was made an Officer of the Order of Canada in 1994. Dr. Thomson was named a Fellow of the Canadian Museums Association in 2000.

Dr. Thomson received her Ph.D. in art history from McGill University in 1981, her M.A. in art history from the University of Maryland in 1974, and her B.A. in history from the University of Western Ontario in 1952. Her particular field of interest is 18th-century French art and architecture. She was appointed Director of the Canada Council for the Arts effective January 1, 1998.

She spoke on several occasions in support of the Millennium Conferences initiative as head of one of the principal partners. The following are notes prepared for her remarks to the Creativity 2000 conference as the introduction to the session entitled "The Creative Mind."

Shirley L. Thomson

Good afternoon, ladies and gentlemen, and on behalf of the Canada Council for the Arts, welcome to the third session of Creativity 2000, Mesdames et Messieurs, permettez qu'en mon nom et en celui du Conseil des Arts du Canada, je vous souhaite la bienvenue; ~ cette troisième séance de Creativité 2000, intitulée « Le cerveau et le processus de création ».

Let me launch our discussion with a few lines of 19th-century verse by Emily Dickinson.

They seem to me as good an introduction as any to this complex subject.

The brain is wider than the sky,
For, put them side by side,
The one the other will include
With ease, and you beside....

The brain is just the weight of God,
For lift them, pound for pound,
And they will differ, if they do,
As syllable from sound.

Creativity has much to do with the complex relationship between the brain, the mind, and the world around us.

Time recently published an article on how brain activity is stimulated by music. It quoted Robert Zatorre, a neuroscientist, as saying: *"We tend to think of music as an art or a cultural attribute, but it is a complex human behaviour ... worthy of scientific study."*

I don't think that this is an either/or situation. Music is an art. Its composition and performance

"Creativity has much to do with the complex relationship between the brain, the mind and the world around us."

are directed through the brain and also through external cultural influences. It may sometimes be associated with a brain pathology.

A recent newspaper article asked, "Was Glenn Gould autistic?" Autism is a little understood and very isolating mental condition. Yet almost all people who suffer from autism are musically proficient and a significant number are highly talented, or even prodigies. Glenn Gould may or may not have been autistic. In either case he was certainly a great interpretive artist, and we appreciate the Goldberg Variations more fully because of him.

Brain imaging technologies are now contributing a scientific explanation of some of the things that philosophers, metaphysicians, and artists have been saying for centuries about the way in which we learn and think. Thales of Miletus argued in the 6th-century B.C. that the essential constituent of all things was water.

Since then, Western philosophers, theologians, and scientists have been exploring the idea that all phenomena can be explained as a chain of effects stemming from a primary cause. This is the central theme of the Enlightenment.

This is not the only possible metaphysical model. The ancient Chinese concentrated rather on holistic properties and harmonious relationships, a line of thinking that is perhaps resurgent again in the environmental debate.

But the search for cause-and-effect relationships has proven remarkably productive in the development of western science. One thing it has done, as Edward O. Wilson[1] has pointed out, is to draw the various branches of science together and demonstrate how they interlock.

Any high school student knows that organisms can be reduced to molecules and understood by the laws of chemistry. And molecules can be reduced to much more elementary particles and understood by the laws of quantum physics.

All the sciences, however, rest upon mathematics, and mathematics derives from human culture, human patterns of thought and belief.

As Northrop Frye pointed out, mathematics *"appears to be a kind of informing or constructive principle in the natural sciences. It continually gives shape and coherence to them without being itself dependent on external proof or evidence."*

"And yet finally the physical or quantitative universe appears to be contained by mathematics.... [There is a] curious similarity in form, for instance, between the units of literature and of mathematics, the metaphor, and the equation."

The conclusion we are led to is that all our knowledge rests ultimately in the human mind's ability to discern pattern.

An Indian mathematician, Ramanujan, born near Madras in 1887 to a poor Brahmin family, possessed one of the most creative minds of the 20th century. His formal mathematical training consisted of what he learned in high school. His mental training, however, included religion: the massively complex, highly evolved pantheism of the Hindu religion; and many of his mathematical insights came to him through religious meditation: he attributed them to the family goddess.

This intuitive and image-laden process we would normally describe as "artistic."

The point is that Ramanujan thought with great freedom, and that this freedom of speculation was rooted in the rich cultural and spiritual texture of his life.

People with sometimes unnerving frequency ask the question, what use is culture and art?

Well, culture can be looked on as a high-speed evolutionary device, an outrunner of biological natural selection, serving the interests of a complex society by bringing into focus what is important to it.

Culture comprises the life of a society, the totality of its religion, myths, art, technology, sports, and all the other systematic knowledge transmitted across generations. It is complex and diverse.

Yet some simple elements of culture — such as the human smile — are so universal that we can consider them genetically determined.

Art says a great deal about how we understand reality. I want to look briefly at the relationship between brain function in the act of seeing and the creation of visual art.

To see, the brain must learn to interpret a flow of changing data in order to grasp essential forms — so that it will be able, for example, to recognize a familiar face as the head turns and the light changes.

Neurobiologists have mapped the way it does this. Interestingly, a person born blind and restored to vision as an adult will find it difficult, if not impossible, to learn to see forms.

Art is concerned with seeing forms at a deeper level than the ordinary seeing of appearances.

Schopenhauer said that painting should strive *"to obtain knowledge of an object, not as a particular thing but as a Platonic ideal, that is, the enduring form of this whole species of things."*

Constable wrote in his Discourses: *"The whole beauty and grandeur of art consists ... in being able to get above ... particularities of every kind ... [to] ... make out an abstract idea of their forms more perfect than any one original."*

Artists as different as Michelangelo and Picasso conveyed their vision of enduring form by leaving a degree of ambiguity as a bridge to the imagination of the viewer. Michelangelo, for example, left many of his sculptures in a slightly unfinished state.

Cubism, the most radical departure in Western art since the introduction of perspective, eliminated of point of view in its search for essential form. It rejected lighting and perspective. It showed figures ambiguously, neither face-on nor in profile. The logical extension of cubism was the rejection of particular objects altogether, and the development of abstraction as, in Mondrian's words, *"the expression of pure reality." "To create pure reality plastically,"* he said, *"it is necessary to reduce natural forms to the constant elements."*

When we perceive an object, as Semir Zeki[2] has pointed out, the brain sees colour first, before form or movement. The straight line is the most basic stimulus to activate a very important category of brain that deals in shape.

A painting such as Barnett Newman's *Voice of Fire* — the most controversial acquisition of the National Gallery during my tenure there as director — stimulates the brain in a specific, immediate, and intense way through the use of strong colours and straight lines.

Voice of Fire, which you can see if you visit the Gallery, is acrylic on canvas. It is a large, tall painting: 5.4 metres high by 2.4 metres wide. On it are painted three vertical columns of colour.

The columns on either side are deep blue with a purplish cast (a combination of prussian blue and ultramarine). The central column is a very warm cadmium red, approaching orange. The contrast is intense. The lines are rigidly straight, although the colours bleed slightly into one another. The texture is flat.

The controversy over its purchase derived, I think, from the fact that abstract expressionism abruptly rejects romantic concepts that underlie many people's ideas of art. Abstract expressionism is assertive and frankly cerebral. It demands an effort of understanding. It shares with mediaeval art the assumption that meaning may lie beyond, rather than in, appearances.

Unlike mediaeval art, however, modern art does not have recourse to familiar symbols to convey its meanings. It is highly subjective.

To quote the Canadian philosopher Charles Taylor: *"unlike previous conceptions of moral sources in nature and God, these modern views give a crucial place to our own inner powers of constructing or transfiguring or interpreting the world."* This internalization corresponds with modern science's emphasis on the brain as the ultimate source of form and meaning.

Newman adds another twist. *"The present painter is concerned,"* he says of himself, *"with the penetration into the world mystery. His imagination is therefore attempting to dig into metaphysical secrets. To that extent his art is concerned with the sublime. It is a religious art, which through symbols will catch the basic truth of life.... The artist tries to wrest truth from the void."*

Digging into metaphysical secrets is the business of both artist and scientist. As the historian of science Giorgio de Santillana wrote: *"So long as it is alive and not sterile, science will remain a speculum entis [a mirror of being]. It will present what metaphysics did, a symbolic structure which is an essential metaphor of being, but is not the only one."*

The creative mind is a mind driven to grapple with these profound questions of form and essence that lie beyond ordinary appearances.

Leaving you with that thought, let me tell you something about the distinguished artists and scientists participating this afternoon.

Catherine Richards is a practising artist and an assistant professor at the University of Ottawa's Department of Visual Arts. Many of you may have seen her interactive installation, "Charged Hearts." It was on view at the media launch for this conference.

I have had the pleasure of knowing Catherine for many years. In 1993 she won the Canada Council's Petro-Canada Prize for her virtual reality work, "Spectral Bodies."

In 1996, when she was artist-in-residence at the National Gallery of Canada and I was director there, the Gallery commissioned "Charged Hearts."

A creative work is creative in two senses. The artist has a creative role as the builder or maker. The viewer also has a creative — and active — role in perceiving and understanding the work.

A great deal of Catherine's work explores that mysterious ground between the physicality of the work of art and the physicality of the viewer, between the scientific phenomenon and the artistic experience. Her current work in progress is about the brain.

"We know we can change heartbeats, and I think we're changing our minds."

Catherine Richards
Visual Artist
Creativity 2000

Dr. Albert J. Aguayo is Director of the Centre for Research in Neuroscience at McGill University. In 1999 he received the Canada Council's prestigious Killam Prize in the Health Sciences. Although most of the Council's prizes are in the arts, the annual Killam Prizes were created in 1981 to honour eminent Canadian scholars and scientists actively engaged in research.

Dr. Aguayo carried out ground-breaking studies at McGill which show that nerve cells in the brain and spinal cord of adult animals retain the ability to regrow and make new functional connections after injury. His research on factors that can promote successful nerve cell growth in the damaged or diseased brain, eye, or spinal cord is important for memory, learning, and other high functions of the mind.

"It seems clear that if we are to have an impact on enhancing the potential for knowledge and creativity of human beings, we have to learn more about how the brain works. I hope also to leave you the idea that by learning about our brains and how to better use them, we may be able to embark in one of the most creative endeavours of all time: understanding ourselves."

"We are moving out of the individual brain and into the collective brain."

"I can tell you that in my own lab, ... it isn't one person who is thinking; it is a group.... It is continuous through new technology.... It is a large community that is thinking and playing with ideas in trying to formulate an answer or design an experiment.

We have a new entity: not only the human brain, but the connected brain.... "

"We are equal, and what makes us different is so little, and yet we have to learn so much about so little because that's what makes us individual and creative."

Dr. Albert Aguayo
Neuroscientist
Creativity 2000

Tedd Robinson was performing at the Canada Dance Festival here in Ottawa last week. I hope that many of you found time to take in his performance and some of the other excellent work presented there.

Quirky originality, humour and pathos, nuance and camp are hallmarks of Tedd's choreography. He achieves a sense of beauty through bizarre juxtapositions, expressing his fascination with both Eastern culture and Western art music. The Council has supported his work extensively.

A choreographer, performer and teacher, he has been commissioned by dance companies and independent artists from coast to coast.

Tedd is based here in Ottawa, where he is the artistic director of *Ten Gates Dancing*, a non-profit organization that promotes the development and performance of contemporary dance works.

Dr. Margaret Boden is an expert on artificial intelligence and the psychology of creativity. Her background includes medical sciences, philosophy and psychology. She was the founding dean of Sussex University's School of Cognitive and Computing Sciences, a pioneering centre for research into intelligence.

As a Professor of Philosophy and Psychology, Dr. Boden lectures widely on artificial intelligence and the human mind. A Fellow of the British Academy, she has published a number of books, among which *Artificial Intelligence and Natural Man* remains a classic.

> *"I don't believe that creativity comes about by magic."*
>
> *"Creativity is part of the natural human world, just as everything else. It isn't in principle beyond the bounds of human understanding, as if it were inspirational from something utterly non-human, or non-cultural."*

Professor Margaret Boden
Creativity 2000

As the 17th-century poet and theologian John Donne said, *"No man is an island, entire of itself. Every man is a piece of the continent, a part of the main."* A vibrant cultural life is a sign of society's health. It also promotes the health of the individual psyche. The creative mind does not develop in a void.

It is fitting that this session on "The Creative Mind" should be sponsored by the **Canadian Institutes of Health Research**, a collaboration of the humanities and the sciences that is working to forge new links in the promotion of a healthy life.

The ability of the **Canada Council for the Arts** to recognize and support artistic creativity is the basis of its whole funding program. The Council made grants of approximately $111 million to artists and arts organizations last year. Roughly 80 per cent of this funding went to organizations and 20 per cent to individual artists.

In an average year, we receive over 15,000 applications for approximately 120 different competitions. All of these applications are –studied by peer assessment committees of professional artists, who must evaluate the creative potential of each project. To carry out our mandate responsibly, we must understand all that we can about the creative process.

Hence our interest in supporting today's conference. I shall be listening intently as this session carries on.

---

[1]Edward O. Wilson is Research Professor and Honorary Curator in Entomology at Harvard University.

[2]Semir Zeki is a professor of Neurobiology and Co-head of the Wellcome Department of Cognitive Neurology at University College, London.

National Science Leader Calls for Support of Arts Education

President
National Research Council Canada
Dr. A.J. Carty

Dr. A.J. Carty has had a rewarding career in science research before being named President of the National Research Council of Canada (NRC).

Dr. Carty spent 27 years at the University of Waterloo where he was successively Professor of Chemistry, Chair of Chemistry, and Dean of Research. Prior to this he spent two years as an assistant professor at Memorial University. Dr. Carty has served on many boards and councils and is currently a member of the Atomic Energy Control Board, the Boards of the Communications Research Centre, the Ottawa–Carleton Research Institute, the Environment Canada R&D Advisory Board, and the Department of National Defence R&D Advisory Board.

Dr. Carty took office as President of the NRC in July 1994. Since then, he has been promoting the vision of NRC as a leader in the development of an innovative, knowledge-based economy through science and technology. He is an active researcher in the field of chemistry and a former President of the Canadian Society for Chemistry. His research interests are in the areas of synthetic chemistry, metal clusters, polynuclear activation of small molecules, and new materials. He has published over 250 papers in research journals, in addition to review articles and book chapters, and has chaired or served on many peer evaluation committees for NSERC and other organizations. He is currently a member of the NSERC Reallocations Committee and the Networks of Centres of Excellence Selection Committee.

Among the numerous honours he has received are the Alcan Award of the Chemical Institute of Canada, the E.W.R. Steacie Award of the Canadian Society for Chemistry, and the Montreal Medal of the Chemical Institute of Canada. He is a Fellow of the Royal Society of Canada and has honorary degrees from the University of Rennes, France, Carleton University, the University of Waterloo, Acadia, and McMaster University.

As President of NRC, he took a very active role in promoting the Millennium Conferences and its themes and establishing new collaborations with the arts and culture communities. As a Canadian science leader, his endorsement of the arts and the importance of arts education brought special credibility to these discussions.

In July 2000, Dr. Carty spoke at the National Art Gallery in Ottawa in the 4th National Symposium on Arts Education. His remarks which reflect upon the Millennium Conferences and Creativity 2000 follow.

Dr. A.J. Carty

I'm very pleased and honoured to open the National Symposium on Arts Education.

Those of you who have come from out of town and perhaps had occasion to wander further down Sussex Drive. About halfway between here and the Governor General's residence, across from the Pearson Building and sandwiched between the British High Commissioner's House and the French Embassy, you come upon an impressive stone building overlooking the river.

This is the National Research Council's original headquarters built in 1932 and still the site where two of our research institutes are located today.

I point this out because, gathered here in this magnificent building, the National Gallery, we're very conscious of Canada's enormous artistic legacy. And down the road, the NRC building reminds us that Canada has a great scientific legacy as well. It's a legacy that the NRC has helped build ever since we were first created in 1916.

In that building, some of Canada's Nobel laureates conducted their research — scientists like Gerhard Herzberg, John Polanyi, Harry Kroto, and Rudolph Marcus. And in that building today, there are young researchers who will be the Nobel laureates of tomorrow.

Out of NRC came some of the science and technology that has created a higher quality of life for Canadians and has made our economy innovative.

Decade after decade, time after time, the NRC has added to Canada's research legacy:

vaccines against meningitis;

electric wheelchairs for quadriplegics;

the first industrial process for the production of magnesium;

the locator beacon that helps rescuers find downed aircraft and the forerunner of the black box;

the heart pacemaker;

the music synthesizer;

synthetic human insulin for diabetics.

These discoveries and inventions that I have just mentioned, as well as a multitude of others, have had a significant impact on society through industrial innovation and wealth generation, health care and technological improvements to our quality of life.

What may not be so well recognized is that many of our innovations have direct application to the arts. In the 1970s, at NRC, researchers invented the algorithms for computer animation and, in collaboration with the National Film Board, created the first animated film *Hunger*. In fact, this milestone in the history of motion pictures was recognized when the two NRC researchers, Marcelli Wein and Nestor Burtnyk, received an Academy Award for their technical creativity in 1998.

This pioneering work at the interface of the Arts & Sciences, while perhaps only a curiosity in the early 70s, has had a tremendous economic impact in the 80s and 90s — the computer animation industry, as illustrated by Steven Speilberg's film *Jurassic Park* and, more recently, *Dinosaur*, is a multi-billion dollars industry, and Canada has continued to be a leader in the field with companies such as ALIAS Wavefront, Soft Image, etc.

Elsewhere at NRC in the 1970s, David Makow, a researcher in the Division of Physics, was pioneering the use of liquid crystals, while his colleague, George Dobrowolski, was exploring the generation and technological applications of thin films. These two researchers began to work together and found ways to use liquid crystals and thin films in painting and sculpture.

They created colours and images that interact with the viewer and the environment. This is an artistic medium for the 21st century and you might be interested in visiting Makow's gallery in Kanata.

> *"Beauty, insight, and invention are augmented by linking the unexpected domains of nature's mysteries."*
>
> **David Makow**
> Scientist, Artist, Inventor

Another medium for the 21st century is virtual reality. NRC has a Virtual Environment Technology Centre in our institute in London, Ontario, and leading edge expertise in virtual reality in our Institute for Information Technology here in Ottawa. Step into the theatre in London, and you could find yourself in the driver's seat of a racecar of the future, or taking a fantastic voyage into the human heart. All in 3D virtual reality of course. Right now, this technology is used by industrial designers, medical researchers, and urban planners.

It is increasingly being put to work in the film world — remember the terrifying scenes of *The Matrix*, a film in which people live in a totally artificial virtual world because their real environment has been destroyed by atomic war. Most certainly, the 21st century will see a dramatic increase in the use of virtual reality by creative artists and society — to carry us into a fantastic, virtual world and to "see" wonders we can only dream of today.

Another art form of the future involves 3D imaging. In the 1980s, NRC pioneered 3D imaging technology. Today, these synchronized 3D laser scanning systems find a vast array of applications from scanning industrial parts for computer-aided design and manufacturing, to the recording of finger and foot prints for forensic sciences; from reverse engineering of silicon chips to the vision system for the international space station.

Museums, including the Museum of Civilization across the river, use NRC's 3D imaging technology to scan paintings, sculptures, and priceless archeological artifacts. The images are digital. You can download them into your computer, manipulate and look at them from any angle in real colour. You can send them around the world over the Internet. You can, in essence, create a virtual museum or art gallery. For a country as geographically diverse as Canada, this opens some exciting possibilities for Canadians to access our national collections from anywhere in the country. It also offers great potential for educational purposes.

NRC has also collaborated with the Canadian Conservation Institute and the National Gallery to capture the full three-dimensional, full colour, images of valuable paintings. The Louvre asked us to scan a collection of paintings by Corot and Rembrandt. The object of this was on the one hand to provide a very detailed picture of texture, colour, and material deterioration — on the other, to provide computer reproductions for display which complement or even substitute for the expensive original.

Last year, another high-resolution laser scanner was taken to Florence to participate in the Digital Michelangelo Project. I am sure that Michelangelo never imagined anything like this being done with his art! But just reflect on what an artist, scientist, and engineer such as Leonardo da Vinci could have achieved with such tools at his fingertips.

NRC is also beginning to recognize and contribute to the application of state of the art technologies for arts education. Two weeks ago, I signed a collaboration agreement with Peter Herrndorf, Director General and CEO of the National Arts Centre. We are working together on tele-education initiatives in the arts.

The NAC's Musical Director, Maestro Pinchas Zuckerman, has long been a champion of distance education. He teaches master classes to violin students around the world via the Internet. We are working with him to use broadband technology so that he can teach these classes in virtual reality. A demonstration of the power of real time interactions of this kind was demonstrated by Maestro Zuckerman at the recent Creativity Conference (which I will tell you about shortly).

I could go on and on about our research, but I think you probably get the point. NRC is at the heart of creativity and innovation in S&T in Canada. Just as this building represents a legacy Canadians have given to the art world, so the NRC building down the street represents the legacy that Canadians have contributed to the world of science and technology.

Last October, when Susan Annis asked me to be the first keynote speaker of your symposium, I wondered, *"Why me? Why would anyone invite a scientist to open a symposium on arts education?"*

There are those who continue to believe that the arts and the sciences represent the opposite ends of a spectrum of human creativity — that if you are drawn to the one, you don't have much interest in the other. But most people who actually are artists or scientists know this is not the case. In fact, there's a very high correlation between artistic and scientific aptitude.

Last year, Canadian researchers examined Einstein's brain. They found that the section of the brain responsible for spatial visioning was uncommonly large. This is the same part of the brain that would be stimulated in a visual artist.

Although best known for his work in theoretical physics, Einstein was also a gifted violinist. In fact, you find many highly qualified people in maths, engineering, and the sciences have musical aptitudes. There's an old joke: what do you get when you bring four Nobel laureates in mathematics together in the same room? Answer: a string quartet.

People in the sciences and in the arts share a similar sense of exploration. They're curious about the world. They want to understand it better — whether the understanding is sought through empirical evidence or through getting in closer touch with the deepest emotions of the human heart. As Albert Einstein expressed it, *"the most beautiful thing we can experience is the mysterious. It is the source of all true art and science."*

People need to express what they have discovered, whether the expression is in the form of an equation or a painting. And I'll say this without any basis in empirical research, but from my own observations — both scientists and artists share a tremendous urge to be creative. And they never lose their sense of childlike wonder.

Einstein once said, *"Imagination is more important than knowledge."* He also pointed out that — *"to raise new questions, new possibilities: to regard old problems from a new angle requires creative imagination and marks a real advance in science."*

Perhaps these words from the *Phantom of the Opera* express it well — and I quote *"close your eyes and surrender to your darkest dreams, purge your thoughts of the life you knew before. Close your eyes, let your spirit start to soar — and you'll live as you never lived before."*

Creativity is as vital to science as it is to art. This is why the National Research Council this year worked together with the National Arts Centre and the Canada Council of the Arts to stage the Millennium Conference on Creativity in the arts and the sciences. This exciting event, which we called "Creativity 2000," took place just 10 days ago at the NAC.

We filled the NAC's theatre with artists, performers, researchers, and mathematicians. Some were professionals. Some were high school and university students. Through the course of the day, they all heard from, or were entertained by, a vast array of creative personalities. Everyone from award-winning novelists to Nobel laureates in chemistry; from violin virtuosos to researchers in neuroscience.

We heard from Douglas Cardinal, the architect of the Museum of Civilization across the river from here, and the artist, Catherine Richards, who is an outstanding innovator in melding the arts and sciences. We linked up by satellite with people like Sir Arthur C. Clarke, author of *2001: A Space Odyssey*, and Benoît Mandelbrot, the father of fractal geometry.

But I was particularly struck by something Douglas Cardinal said. In his words, *"We are magical beings, as human beings, because we have this power of creativity. Creativity is creating something from nothing. So, entering the creative domain you must be willing to stand on what is known and leap into the vast abyss of the unknown. The known universe has already been created.*

The creative world is the vast abyss beyond what is known. It is a land of total possibility, a blank sheet of paper that is the land of the eagle, where our true power as a human being 'dwells,' is found, and is expressed."

It was a wonderfully stimulating day. I think we proved conclusively that creative people in both the arts and the sciences have a great deal to offer one another — both in their insights and the products of their imagination.

The audience had an opportunity to raise questions and discuss issues with the guest speakers. One of the most compelling questions came from a high school student. She described her difficulty in choosing a career path. She was very involved in her school drama program, but she was also an honours student in maths and sciences. Her peers and her elders were urging her to make a career choice. They told her that, in this very competitive marketplace, you must choose a niche. Do it quickly! Do it young! Build a career in one field of study. But she didn't want to abandon her love of the theatre. Or her love for mathematics, chemistry, physics, or biology. What was she to do?

She raised this question with a panel that included an architect, a famous chemist, and a movie writer/director/actor. The consensus was that the important thing was not to lose her passion for the things that interested her. They urged her to maintain the breadth of her interests. And it seems to me that, out of the range of interest and talent of this young woman, something unique may evolve. Something that those urging her to make a choice may not have thought of as a career option.

But for this audience, and this symposium, her dilemma raises interesting issues. Canada needs more men and women who cross traditional boundaries and excel in both arts and sciences. We should discourage the bifurcation between the arts and the sciences that occurs at a relatively early age and positively encourage students to develop significant talents in many disciplines.

Indeed one of the recurring themes of the conference was the importance of early childhood education in stimulating the brain's synapses. This is a powerful theme among educators (and scientists) in Canada, including Dr. Fraser Mustard who chaired a 1999 study on early childhood education for the government of Ontario. A key finding in his report called *The Early Years — Reversing the Real Brain Drain* was that during the early years of a child's life, the billions of cells in the brain are poised to make connections.

This "hard wiring" of the brain occurs through stimulation of specific parts — for example, through reading, through the challenge of learning a foreign language, or the stimulus of an early exposure to science.

Most importantly, neuroscience and human development studies show that this early wiring and sculpting of the brain influences learning, behaviour, and health status throughout life. In fact, scientific research shows that the nurturing a child receives during the first three years of life is critical to the healthy development of the brain.

If I can quote Fraser Mustard here — *"Stimulation and interaction help to activate and 'wire' the connections between brain cells. Experiences during a baby's first days, months, and years affect brain structure which has an impact that lasts a lifetime."*

One of the most studied areas is language, and we are at last beginning to accept that the very young have an incredible ability to pick up new languages — they have the capacity to function truly like sponges — they soak up knowledge. This is an ability that disappears around age 12. I am sure that this city is filled with 50-year-old bureaucrats who can testify to the difficulty of trying to learn a second language as an adult.

But if we can lay down a foundation for languages at a very early age, why not a similar grounding in the sciences and the arts. Each discipline has its own "language" if you will, and introducing the full range of possibilities cannot help but open up an enormous potential for the future.

At our conference last week, I had a very interesting conversation with a colleague from Finland who told me that in his country, children at the earliest age are all taught a second language, learn to play a musical instrument, and must study math and science — all before what we would consider grade one. This is a visionary approach that I think Canada would do well to emulate.

Our children are capable of understanding much more than we give them credit for. And at these early ages, the exploration of new ideas, new modes of expression, new languages, can't help but be fun. We should teach our children to enjoy creativity; enjoy exploration, enjoy the physical world around them. Once we have established that early link between learning and fun, we have a foundation on which an individual can build for the rest of his or her life.

Ladies and Gentlemen, I don't think I need to preach to the converted here. I think we share a commitment to early childhood education and to the progressive evolution of our educational system. I also believe, that we all share an enthusiasm for imagination, creativity, innovation, and expression — both for the arts and the sciences.

But in helping you to launch the Fourth National Symposium on Arts Education, I want to assure you that the world of science and technology takes a great interest in the issues you are wrestling with.

We live in an age of convergence. We find convergence everywhere — certainly in science and technology. Computers and telecommunications technologies converged to create the Internet. At NRC, we're converging technologies as disparate as organic chemistry and electronics.

And it is my prediction that, in the coming years, the worlds of the arts and the sciences will begin to converge as well. Scientific research, the performing arts, the written and visual arts, may seem as far apart as Einstein and Wagner. But let us not forget that in the past some individuals have superbly transcended science and art.

Just think of Leonardo da Vinci, an architect and sculptor of the first order, an artist, a scientist and an engineer — a man whose creative instincts lead him in many directions — and the world reveres his contributions. Why should we not aspire to "free" our children to pursue a diversity of interests, with the sciences and the arts and humanities no longer alternate choices but symbiotic and desirable partners.

At the beginning of a new millennium, we find ourselves in a world where our quality of life is shaped by innovation. We talk about a "knowledge-based economy." But, really, the application of knowledge is much more profound. It affects everything we do: from the quality of our health care to how we entertain ourselves; from how we communicate with one another to how we transact business.

The arts — whether visual or performing, whether the written word or the creation of edifices like this wonderful gallery — have a way of opening the mind. They stimulate the synapses, and make one more receptive to creativity.

"... I want my imagination rekindled. I want to be shown how to look at things in new ways. And I believe my capacity for innovation and creativity in my own discipline will grow as a result."

They send charges down different circuits of the brain, and open new possibilities. As a scientist, I want my imagination rekindled. I want to be shown how to look at things in new ways. And I believe my capacity for innovation and creativity in my own discipline will grow as a result.

That's one good reason for promoting arts education. There are many, many more.

But at a time in our history when Canada is racing to keep at the forefront of knowledge and innovation, I would say that this reason alone should make every policy maker a champion of the arts.

Nobel Laureate Said Embrace Arts and Sciences for Their Own Sakes and Extend Boundaries of Knowledge

Gerhard Herzberg
1904–1999

In March 1999, as plans were being finalized for the beginning of the Millennium Conferences on Creativity in the Arts and Sciences, Dr. Gerhard Herzberg, a Canadian physicist and Nobel laureate passed away. He spent his life in the pursuit of science but also loved and found time for the arts, especially music.

Several events were dedicated to the memory of Dr. Herzberg. These included a performance by the Canadian tenor Ben Heppner with the National Youth Orchestra of Canada and the National Arts Centre Orchestra which was conducted by Mario Bernardi on July 28, 1999. On that occasion, Dr. David Leighton, Chairman of the Board of Trustees of the National Arts Centre said Dr. Herzberg had been "a remarkable Canadian" whose "passion for the arts and sciences is an inspiration to all of us." From October 30 to November 3, 1999, a conference titled "Inspired by Herzberg: Spectroscopy for the Year 2000" was held in Cornwall, Ontario. This was a continuation of conferences that began when Dr. Herzberg was sixty-five and were held every five years. Further, the series of Conferences on Statistics, Science and Public Policy held at Herstmonceux Castle in the United Kingdom which have sought to encourage interaction among disciplines are part of Dr. Herzberg's legacy. In April 2000, the association between that conference series and the Millennium Conference was forged when the latter supported the participation of Miss Angela Hewitt, the renowned Canadian pianist, in a celebration of the new alliance between the arts and sciences.

Although Dr. Gerhard Herzberg did not live to witness the Millennium Conferences on Creativity in the Arts and Sciences, his spirit was felt throughout, and in many ways, his life exemplified the ideals promoted by the series. Some of his thoughts and the example of his life are captured in the following essay written by his daughter Dr. Agnes M. Herzberg. Dr. A.M. Herzberg, who attended Creativity 2000 as well as events specifically dedicated to her father's memory, is a Professor in the Department of Mathematics and Statistics at Queen's University.

Dr. Gerhard Herzberg

Dr. Herzberg's advice, his science, his work on behalf of dissidents, his campaigns for the funding of pure science by government, and his sense of fun will always be remembered through the millennium conferences.

Born on Christmas day in 1904 in Hamburg, Germany, Gerhard Herzberg had early on wanted to be an astronomer, but this was impossible as he lacked private means. He was also interested in physics and managed to obtain a private scholarship for the first two years at the Technical University in Darmstadt, and over the following years, he received public funding. After his graduation from the Technical University with a Doctorate in Engineering (Dr.Ing.), he held post-doctorate fellowships at the University of Göttingen and the University of Bristol. He then returned to a position at the Technical University in Darmstadt.

In 1933, Dr. J.W.T. Spinks, a young chemistry professor at the University of Saskatchewan, came to Darmstadt to spend a year working with Dr. Herzberg. Because the political situation in Germany was deteriorating, Dr. Spinks managed with the help of the President of the University of Saskatchewan to obtain a position for Dr. Herzberg at the University; the Carnegie Foundation of New York provided the initial funding. Dr. Herzberg said that his experiences at the University of Saskatchewan and in Saskatoon, from 1935 to 1945, were the best ten years of his life. From that time on, he identified himself as a Canadian.

"A high standard of living is not, as such, a goal worth striving for unless a high standard of living includes a high standard of art, literature, and science ..."

After three years at the Yerkes Observatory of the University of Chicago in Williams' Bay, Wisconsin, Dr. Herzberg returned to Canada in 1948, to the National Research Council of Canada. He was appointed Director of the Division of Physics in 1949 and retired from that position in 1969. He remained at the Council as Distinguished Research Scientist and continued his research for almost the next thirty years until illness made this impossible.

Dr. Herzberg was awarded many honours, including the Nobel Prize for Chemistry in 1971. He did not forget the University of Saskatchewan:

"It is obvious that the work that has earned me the Nobel Prize was not done without a great deal of help. First of all, while at the University of Saskatchewan, I had the full and understanding support of successive Presidents and of the Faculty of the University who, under very stringent conditions, did their utmost to make it possible for me to proceed with my scientific work."

Before taking on the position of Director of Physics, he was told that his *"main administrative function would be to find first-class people for the positions that were open and let them use their own judgement in selecting their research projects rather than try to direct them."*

Dr. Herzberg believed that advances in science stem from research done when individuals are given free rein and not restricted by the policy decisions of government agencies or committees. Demonstrating his belief that administrators are meant to support science, Dr. Herzberg wore a white laboratory coat even when confined to his desk. The key to scientific success and technological advance, he felt, was in the laboratory, and not in the office or committee room.

He also defended publicly his conviction that a society should maintain a high standard of progress in science and the arts; survival should not be society's only goal. He felt that all citizens needed to consider *"the works of art, literature, and basic science as not merely the icing on the cake but as the essence of human existence."*

Dr. Herzberg worried about the divide between the arts and sciences. In a paper, "Remarks on the Boundaries of Knowledge," he wrote:

"The limitations to the advance of knowledge that are introduced by the limited extent of our memory (and I realize that there are many other reasons) are illustrated by the sharp schism between literary intellectuals on the one hand and natural scientists on the other, so eloquently described by C.P. Snow in his little book The Two Cultures and the Scientific Revolution. *Of course, this schism really goes much further, as was recognized by C.P. Snow; even among scientists of one discipline there is considerable lack of understanding of the work in other disciplines. While much of this schism is caused by the limitations of the average intellect, the division is further fostered to a considerable degree by our educational system, which in my opinion gives far too much choice to students of our high schools."*

He went on,

"... let us develop a cultural climate which believes that human excellence is a good thing in itself, a climate in which all members of society can rejoice and delight in the things that the small number of exceptional members is able to do, without asking what use they have for survival. ... Without that, to quote C.P. Snow, 'some of the major hopes, the major glories of the human race will rapidly disappear.'

... the striving for excellence in order to increase our cultural heritage and to extend the frontiers of knowledge is the most important aim of mankind. There may be more urgent things to be done in connection with our survival. But we must not let these necessities detract us entirely from our devotion to the striving for knowledge for its own sake, to the aim of understanding man and his world."

In summary, the words Dr. Herzberg used some time ago are appropriate not only for Canada, but for all countries:

"A high standard of living is not, as such, a goal worth striving for unless a high standard of living includes a high standard of art, literature, and science.

... If Canada is to be economically prosperous without at the same time supporting the arts and sciences for their own sakes, it will not reach the level of a great nation."

Call for a National Forum on Creativity and Innovation

In late May 2000, the Symposium on Creativity and Innovation in the Arts and Sciences was held as a featured part of the 2000 Congress of the Social Sciences and Humanities at the University of Alberta in Edmonton in association with the Millennium Conferences.

A compilation of suggestions made at the symposium has been prepared as a special report for the Honourable Herb Gray, P.C., M.P., Deputy Prime Minister, and Minister responsible for the Millennium Bureau of Canada. This report constitutes a legacy of Canada's national Millennium celebrations from the partners, the Humanities and Social Sciences Federation of Canada (HSSFC), the Social Sciences and Humanities Research Council (SSHRC), the National Research Council (NRC), and the University of Alberta. (Appendix I).

The chair has called for a National Forum or Task Force on Creativity, Inventiveness, and Innovation *"to explore and identify ways in which new and original ideas can be encouraged and developed across the full spectrum of the arts, sciences, medicine, social sciences, business, industry, and technology in Canada for the use and benefit of all Canadians."*

D.M.R. Bentley
Professor
Department of English
University of Western Ontario
and Chair of the Symposium
on Creativity and Innovation in
the Arts and Sciences

Appendices

Appendix I. Report of the Symposium on Creativity and Innovation

Humanities and Social Sciences
Federation of Canada

Report based on the Symposium on Creativity and Innovation in the Arts and Sciences
Edmonton, Alberta
May 25–26, 2000

Expert Participants

Dr. Sharon Bailin
Faculty of Education
Simon Fraser University
Burnaby, British Columbia

Dr. Cheryl Bartlett
School of Science & Technology
University College of Cape Breton
Sydney, Nova Scotia

Dr. Min Basadur
Michael G. DeGroote School of Business
McMaster University
Hamilton, Ontario

Dr. David Bentley
Department of English
University of Western Ontario
London, Ontario

John Bonnett, Ph.D. (cand.)
Guest Worker
Interactive Media Research Laboratory
NRC
Ottawa, Ontario

Dr. Tom Brzustowski
President
NSERC
Ottawa, Ontario

Professor William Buxton
Chief Scientist
Alias | Wavefront Inc.
Toronto, Ontario

Mr. Douglas J. Cardinal
Douglas J. Cardinal Architect Ltd.
331 Somerset Street West
Ottawa, Ontario

Dr. Patricia Demers
Vice-President
SSHRC
Ottawa, Ontario

Ms. Sara Diamond
Artistic Director MVA
The Banff Centre for the Arts
Media & Visual Arts
Banff, Alberta

Dr. Jim Dueck
Assistant Deputy Minister
System Improvement and Reporting (SIR)
Edmonton, Alberta

The Honourable Herb Gray, P.C., M.P.
Parliament Hill
House of Commons
Ottawa, Ontario

Dr. Peter A. Hackett
Vice-President, Research
NRC
Ottawa, Ontario

Mr. Keith Kelly
Director
Public Affairs, Research and Communications
The Canada Council for the Arts
Ottawa, Ontario

Dr. Sonia Mansour-Robaey
UQAM
Montreal, Quebec

Dr. Michael J. Murphy
Academic Director
Rogers Communications Centre
Ryerson Polytechnic University
Toronto, Ontario

Dr. Joe Norris
Department of Secondary Education
University of Alberta
Edmonton, Alberta

Dr. Tanya Prochazka
Department of Music
University of Alberta
Edmonton, Alberta

Dr. Robert Root-Bernstein
Department of Physiology
Michigan State University
East Lansing, Michigan

Dr. Mark A. Runco
Department of Child and Adolescent Studies
California State University, Fullerton
Fullerton, California

Professor Marc Spooner
Faculty of Education
University of Ottawa
Ottawa, Ontario

Dr. Marjorie Stone
Assistant Dean (Research)
Faculty of Arts and Social Sciences
Dalhousie University
Halifax, Nova Scotia

Dr. Louis P. Visentin
President
University of Brandon
Brandon, Manitoba

Mr. Rudy Wiebe
8510-104th Street
Edmonton, Alberta

Preface

For several years now, it has been apparent to many observers that if the economic, social, and even physical well-being of Canadians is to be enhanced in the 21st century, Canada's governments, educational institutions, and granting agencies must work in concert to foster and support the creativity and innovation upon which success increasingly depends in the brave new world of DVDs, Auto-CAD, bioengineering, and a host of other recent and emergent technologies and disciplines. Such, in any event, was the perception that prompted me in November 1998 to write to the Chair of the Canada Council, the President of the Humanities and Social Sciences Federation, the presidents of the federal granting agencies, and the Deputy Prime Minister, the Honourable Herb Gray, to urge the creation of "a Task Force on Creativity, Inventiveness, and Innovation whose mandate would be to explore and identify ways in which new and original ideas can be encouraged and developed across the full spectrum of the arts, sciences, medicine, social sciences, business, industry, and technology in Canada for the use and benefit of all Canadians." The Symposium on Creativity and Innovation in the Arts and Sciences of which the present Report is a product was the result of that suggestion, and a testament to the generous support of the HSSFC, the Social Sciences and Humanities Research Council, the National Research Council, and the University of Alberta. The Symposium's greatest debt, then, is to Louise Forsythe, Marc Renaud, Arthur Carty, Peter Hackett, Garth Williams, Paul Ledwell, and others at the HSSFC, SSHRCC, the NRC, and the University at Alberta whose generosity and enthusiasm brought it to fruition at the Congress of the Social Sciences and Humanities in Edmonton on May 25 and 26, 2000. That the Symposium was the success to which this Report testifies was due to the technicians who made many portions of it possible, to the participants who made every moment of it both instructive and delightful, and, last but by no means least, to its principal organizer and amanuensis, the indefatigable and imperturbable Marc Spooner.

Whatever else they may be, creativity and innovation are processes that bring useful and valuable ideas and things into existence. As this Report attests, the Symposium on Creativity and Innovation in the Arts and Sciences was part of such a process — an occurrence (and, occasionally, a happening) in a series of actions and events whose future outcomes, it is greatly to be hoped, will be of use and value to all Canadians. Of course, the Report, so ably assembled by Professor Spooner, is a further and crucial occurrence in that series — crucial because it is the major stage in generating radical thinking and radical action by those who can assist in decisive and constructive ways in fostering and supporting creativity and innovation in Canada. Thus a final debt of gratitude must go to the Honourable Herb Gray, who not only travelled to Edmonton to address the Symposium with his characteristic wit and wisdom, but has also undertaken to ensure that the substance and recommendations of this Report will be given serious consideration by the Federal Government.

D.M.R. Bentley, FRSC
Professor
Department of English
University of Western Ontario
Waterloo, Ontario

Coordinator's Message

As coordinator of the Symposium on Creativity and Innovation in the Arts and Sciences, I am pleased to submit the following Report, which represents highlights and themes that emerged from the presentations and discussions that took place in the course of the two-day Symposium at the University of Alberta on May 25 and 26, 2000. The Report does not attempt to do justice to the wide array of topics centred on the creativity theme that emerged at the Symposium; rather, it aims to provide a brief and informative overview of several of the important issues that arose, particularly those that relate to the role of education, Granting Councils, and the Federal Government in fostering and supporting creativity and innovation. Numerous other promising directions of enquiry and many issues of greater specificity also emerged from the Symposium but are necessarily mentioned only in passing in this Report.

The recommendations flowing from the Symposium can be viewed as a consensus for increased participation at every level of education and government in the promotion of creativity and innovation as crucial sources of the future wealth and standard of living of all Canadians. If Canada is to continue to move forward, creativity and innovation must be recognized as key components of Canadian society in the 21st century.

The following Report is based on a collection of information sources that include personal and rapporteur notes, audio recordings of sessions, speakers' notes, as well as scholarly publications and Provincial and Federal Government reports.

Yours sincerely,

Marc Spooner
Coordinator, Symposium on Creativity and Innovation in the Arts and Sciences
Edmonton, Alberta
May 25–26, 2000

Acknowledgements

Both personally and on behalf of the Humanities and Social Sciences Federation of Canada (HSSFC), the Social Sciences and Research Council of Canada (SSHRC), and the National Research Council of Canada (NRC), I would like to thank all the participants in the Symposium on Creativity and Innovation in the Arts and Sciences held during the 2000 Congress of the Social Sciences and Humanities at the University of Alberta in Edmonton on May 25 and 26, 2000.

The Symposium exceeded its goal of bringing together in constructive colloquy researchers from across Canada and beyond in the natural, applied, and social sciences as well as the arts and humanities to discuss the topic of creativity and innovation. The inspiring spirit of collaboration among the disciplines could be readily and continuously felt as presenters and audiences alike explored suggestions and solutions for promoting creativity and innovation in Canadian society.

That the Symposium was a success on so many levels was due in great measure to its participants, all of whom deserve thanks for their time, effort, insights, and wisdom. Once again, a heartfelt thank you for your interest and effort in this regard. A special note of thanks goes out to Professor David Bentley, Garth Williams, Paul Ledwell, and the University of Alberta for helping to bring this idea to fruition.

Marc Spooner
Part-time Professor
Department of Education
University of Ottawa
Ottawa, Ontario

Table of Contents

Executive Summary

"No country can afford to neglect its creative minds."

Dr. Frederick Banting
(as cited in Jackson, 1976, p. 199)

Canadians realize that innovation is vital. It is the foundation of our economic and social prosperity ... To be successful, we need a steady stream of new ideas...

Sustaining Canada as an Innovative Society: An Action Agenda, 1997, p. 5

It is widely agreed that if Canada is to continue to develop and prosper in the 21st century, creativity and innovation are of key importance. The health and wealth of organizations, cultures, and nations are rooted in human capital and the generation of productive ideas. In order to maintain and improve the Canadian quality of life both individually and collectively, the generation and promotion of such ideas by Canadians must be encouraged at every level: individually, scholastically, organizationally, and culturally.

Focus

This Report is focussed on themes and recommendations that emerged during the two-day Symposium on Creativity and Innovation in the Arts and Sciences Symposium held in conjunction with the 2000 Congress of the Social Sciences and Humanities, University of Alberta, Edmonton, May 25, 26, 2000.

Defining Creativity:

Creativity is imaginative, motivated, transformative, and productive thinking and activity within a particular context or framework of knowledge, inquiry, and skills — a process that generates outcomes which are original, significant, effective, and of value or use (or both) to the community.

Definition of creativity generated by the question "What is creativity?" posed during the opening session of the Symposium

Main Directions

The most wealthy nations and regions of the world are the most innovative;

Most new jobs depend on innovation;

Innovation is a renewable source of wealth, and the status quo is not a viable option.

creating an innovation culture: Key Challenges and Opportunities as Ontario Moves Ahead in the New Millennium, 1998, p. 3

What follows is a blueprint for the direction in which Canada needs to advance if it is to continue to compete successfully in a knowledge-based economy. The aim of this Report is to increase the economic prosperity and quality of life of all Canadians. More specifically and most important, it is a prospectus for fostering creativity and innovation as vital components of a vibrant, healthy, and prosperous people and nation. Canada must continue to enhance its position as a nation where creativity and self-actualization, originality, and innovation are openly encouraged, fostered, and rewarded.

Emergent Themes

1.0 Creativity: The Process

Creativity is a process that can be taught and learned by all. Creativity should be viewed as a process, a way of thinking, not exclusively an act, an event or a product. It is a learned skill, not an innate trait or gift that some people possess and that others do not. Creativity requires tools and thinking patterns that are convergent as well as divergent, logical as well as intuitive, and evaluative as well as generative.

Creativity is frequently an "effective surprise" but it is also the painstaking result of trial and error.

When we speak of creativity and innovation, we are not simply discussing the introduction of new goods or services to the market. Moreover, focussed, targeted problem-solving is typically not activity of the most creative and innovative kind.

2.0 The Value of Art–Science Interaction and the Benefits of Multi-disciplinary Study and Teamwork

The best artists and scientists combine elements, skills, and tools from a variety of sources. In order to foster a culture of creativity, Canadian governments and schools must actively promote a culturally literate populace, capable of integrating and synthesising various forms of knowledge-creation.

"History shows that the sciences and technology have never flourished in the absence of a similar flourishing of the arts" (Root-Bernstein, 1997, p. 15). Renaissance Italy is a case in point, as are nineteenth-century Britain and France.

3.0 Recommendations for Governments, Schools, and Granting Councils

3.1 Encourage Multi-disciplinarity

a) Schools should continue to encourage an integrated curriculum with assignments and classroom activities that are problem — rather than discipline-based.

b) Universities should promote cross-disciplinary study, for example, allowing and encouraging students to combine seemingly unrelated fields such as music and engineering or computer science, history, and education.

c) The Federal Government should make additional funds available to the Granting Councils that are specifically earmarked for joint council sponsorship of truly cross-disciplinary initiatives — not just teams from closely allied disciplines.

3.2 Risk, Play, and Great Expectations

d) Granting Councils should allow for numerous small research grants to be allocated to projects that are seen as high-risk, highly likely to fail, but nonetheless low-risk in terms of monetary investment and potentially high-gain in terms of creativity and innovation.

3.3 Time to Explore and a Tolerance for Ambiguity

e) Granting Councils should also be willing to tolerate basic research that flows in unspecified directions, is flexible in nature, and that follows ad hoc and intuitive ventures into unexpected areas.

3.4 Individual and Organizational Re-invention

f) In order to push the boundaries of excellence, organizations should remain fluid enough to allow for the rapid formation and dissolution of teams that find themselves excited by particular problems.

g) Universities should be flexible, perhaps allowing a certain percentage (5%) of any curriculum to be comprised of elective courses that bypass standard assessment procedures typically involved when attempting to introduce new courses. Programs should be reevaluated and possibly redesigned every five years.

3.5 Creating a Culture of Creativity and Innovation

h) Education conceived as a critical and dialectical process (Popper, 1994) between the individuality of the child and a cumulative and diversified knowledge is the best means to foster creativity in a society.

i) If the Federal Government truly wants to create a culture of creativity and innovation, then it must also encourage and make additional funds available for exchanges between researchers from diverse cultures.

j) Granting Councils should reward researchers who have been successfully creative and innovative in the past, even if they are now looking to challenge their creative minds by attempting to work in new fields.

k) Governments should be sponsoring colloquia, debates, and public lectures on the topic of creativity and innovation in the Arts and Sciences.

l) The general public must be made aware of the increasingly crucial importance that creativity and innovation are playing in the knowledge-based economies of the 21st century.

m) A greater awareness and respect for the cross-pollinary effects and seminal importance of Science–Art interactions must also be generated, for, as observed earlier, science and technology have never flourished in the absence of a similar flourishing of the arts.

n) Professor Bentley's call for the formulation of a Task Force on Creativity, Inventiveness, Innovation, and Culture, falling under the combined aegis of the Granting Councils (Canada Council, CIHR, NSERC, and SSHRC), the Humanities and Social Sciences Federation of Canada (HSSFC), the National Research Council (NRC), as well as other interested agencies and organizations, should be acted upon.

Part A: Creativity: The Process

Creativity research is typically undertaken from the perspective of either the person (traits), the product, the process, or the press (environment). However, in the past an overemphasis has been placed on the person (traits) and the products resulting from creative endeavours. It is now largely agreed that it is far more useful to focus on creativity as a process and to illuminate environments that foster this process. But first, given the nebulous nature of creativity as a concept, assumptions and positions must be clarified, folklore and myth debunked.

Creativity can be fostered and taught. Creativity, as a process, is a way of thinking. It is not exclusively an act, an event, or a product. It is not necessarily an innate trait or gift that some people possess and that others do not (though a disposition towards creativity may be partly predetermined), but rather it is a learned skill. It should be noted that creativity is not a characteristic of persons irrespective of specific achievements. That a person may exhibit certain personality traits or cognitive attributes — for example, fluency, flexibility, spontaneity, risk-taking, and divergent thinking — may be helpful but they are not on their own enough to constitute a creative person.

Creativity requires tools and thinking patterns that are convergent as well as divergent, logical as well as intuitive, and evaluative as well as generative. Contrary to popular belief, creativity is not the result of a radical break with tradition; rather, it seems that creative offerings lie on a continuum — in one direction, an originality that occurs within the constraints of a given tradition, and in the other, an originality that involves an alteration of some aspects of the constraints themselves. Paradigms shift when a critical number of a tradition's constraints are altered (Kuhn, 1996). In other words, known concepts are either extended into new areas of application, or in the highest-order creativity, altered is the universe of meaning itself (Ghiselin, 1963).

1.1 Recognizing the Problem

Creativity is a process that involves the invention or recognition of a problem, the solutions to which are likely to be surprises to both the creator and to his or her peers. Creativity is, as Jerome Bruner has observed, an "effective surprise." It is the painstaking and wondrous result of trial *and* error, and a comfort or tolerance for ambiguity.

As the Nobel prize-winning chemist Sir Harry Kroto states, "nine out of ten of my experiments fail, and that is considered a pretty good record amongst scientists"

All Our Futures: Creativity, Culture and Education, 1999, p. 34

Creativity is the purview of perseverance with "chance favouring only the prepared mind," a slogan Louis Pasteur was fond of stating.

As cited in Hayes, 1989, p. 136

1.2 Innovation: More Than New Goods and Services

Let us be clear that when we speak of creativity and innovation, we are not simply discussing the introduction of new goods or services to the market. Rather, as stated in *Closing the Gap: Investing in Knowledge for a Better Canada* (1998), "to promote economic growth and the well-being of Canadians, we must look beyond the customary approach to innovation that focusses solely on technology-based innovation" (p. 8).

Technology-based innovation may be one component of innovation, and rightfully an important one; however, creativity that is far-reaching, that leads to transformative innovations, is creativity that is curiosity-driven and frequently serendipitous. Focussing on targeted problem-solving is typically not research activity of the most creative kind.

Studies of innovation in the biomedical sciences show that over fifty percent of biomedical breakthroughs occur from projects funded for other purposes... By the time something is fundable, it is long past being creative. Funded projects are those that fill in acknowledged gaps using acceptable methods. Surprises are not expected nor desired.

Root-Bernstein, 2000, p. 9

It is crucial that governments, schools, and Granting Councils continue to realize that, although projects may be discipline-based, problems and their solutions are not.

But how can organizations, schools, and Granting Councils develop environments that encourage the creative process to flourish?

Developing and enhancing the creative process, and fostering environments conducive to creative enquiry and innovation will be the focus of **Part B** of this Report.

Part B: Enhancing Creativity

2.0 The Value of Art–Science Interaction and the Benefits of Multi-disciplinary Study and Teamwork

Creative thinking is facilitated in individuals who possess complete, well-rounded, and complementary forms of literacy and/or numeracy. The pathway to understanding is discovered using complex and inter-related human tools, intelligences, perceptions, and sensibilities. The best artists and scientists combine knowledge, skills, and tools from a variety of sources.

> *...many scientists employ the arts as scientific tools. Moreover, various artistic insights have actually preceded and made possible subsequent scientific discoveries. The arts thus can stimulate scientific progress, and we dismiss them at our peril.*
>
> *History shows that the sciences and technology have never flourished in the absence of similar flourishing of the arts. The reasons for this connectedness have become apparent in the past several decades... A consensus is emerging that scientists and engineers need skills associated with, and often learned from, the arts.*
>
> *Clearly, the arts are not the useless, frivolous, or purely subjective pastimes they are often portrayed as being. My thesis is therefore very simple: If we let the arts atrophy in this country through lack of public support, we also will lose an important part of the creative base from which the next generation of scientific and engineering breakthroughs will emerge.*
>
> Root-Bernstein, 1997, p. B6

In order to foster a culture of creativity and innovation, it is of foundational importance that all citizens are given a full opportunity to achieve a high level of literacy and numeracy. In the new knowledge-based economy, governments must encourage a culturally literate populace, capable of integrating, synthesising, and advancing various forms of knowledge creation.

> *What does fostering Canadian industry's economic competitiveness have to do with history, sociology, English literature, or any of the other many disciplines that comprise the social sciences and humanities? The reality is, a lot. As 30 CEOs of Canadian high-tech companies recently said in a public letter, they need more than merely technical training. They need people who are trained in the social sciences and humanities as well.... Not only must we make wise investments across the entire spectrum of disciplines and fields of knowledge in both the physical and social sciences but just as critical is the creation and maintenance of links between them. Over the past several years, there has been much study on the question of whether the humanities and social sciences have a place in the new economy — and the answer has been a resounding YES. The*

reasons are not hard to find. In a knowledge-based economy, the skills acquired in a social sciences or humanities education are relevant, readily adaptable, and indicate a searching intellect. Businesses competing in a global economy need ... people who are able to discern emerging patterns of behaviour. They need people who know how to find answers, or where to look for information that will help them draw their own conclusions. They need people capable of lifelong learning, and who address new challenges as opportunities to explore new ideas and develop new skills.

The Honourable Herb Gray, P.C., M.P.
Edmonton Symposium

Given the fragmentary nature of current knowledge, the sort of comprehensive knowledge idealized in the "Renaissance person" may no longer be an attainable reality; however, the Renaissance team is most certainly achievable. Research teams comprised of members from a variety of disciplines have a greater likelihood of developing innovative solutions to given problems.

How does encouraging multi-disciplinary study and teamwork work?

Keeping in mind that creativity is a process, it may be counterintuitive to conceptualize this process as identical among various domains and disciplines. Nevertheless, the variety of *tools* that facilitate creativity and innovation and that are employed by the world's most creative people may be taken collectively as a common set. For example, artists, scientists, researchers, educators, and inventors share in various combinations and degrees the same cognitive and intuitive tools at the core of creative understanding; however, the manner in which they utilize these tools and the creative products that result are manifest in substantially different ways, depending on their disciplines of choice. Within disciplines, specific styles of thinking tend to be emphasized at the expense of others. At the same time, each thinking style and/or tool is equally valuable for creative understanding, and together as a set operate synergistically.

2.1 The Creativity Toolbox

Metaphors, empathetic thinking, and imagining are often useful in pattern-breaking and seeing in new ways. Multi-disciplinary work facilitates the discovery of preconceptions and habits of thought, and this, in turn, leads to the extension, modification, and variation of thinking styles. Multi-disciplinary collaborations promote new ways of observing and create an awareness of old patterns as well as new ones. Multi-disciplinary teams and training facilitate seeing not only what other's have seen, but also what they have missed, and thus learning to see in new ways.

The Wittes and Kerwin have defined four basic types of ignorance.... The major ones are: 1) the things we know we do not know (overt ignorance), which are the things that normal disciplinary research tends to address; 2) the things we don't know we don't know (hidden ignorance), which are things we discover through error, serendipity, or trans-disciplinary comparisons; 3) the things we think we know but we don't (ignorance masquerading as knowledge), which usually survive because of untested assumptions or inadequate skepticism; and 4) the things we think we do not know but really do (knowledge masquerading as ignorance), which often takes the form of nontraditional sources of information ignored by existing disciplinary paradigms.

Root-Bernstein, 2000, p. 12

Effective scientists bring empathy, emotion, imagination, and a sense of play into their research projects and their classrooms. They frequently use metaphor, visualization, and real-life examples to ensure richer understanding of the problem or subject matter at hand. As disciplinary metaphors and forms of envisaging vary, so too do perspective and insight.

In fact, when the world's most creative thinkers are questioned, we find a basic set of tools that in various combinations are at the root of their creative understanding and that guide their creative endeavours. These tools are not the sparks that arise mysteriously from the mind of genius; on the contrary, they are tools that are known and may be learned and honed. They include: *observing, imaging, abstracting, recognizing patterns, forming patterns, analogizing, body thinking, empathizing, dimensional thinking, modelling, playing, transforming, and synthesizing* (Root-Bernstein & Root-Bernstein, 1999; see pp. 25–27 for definitions). Each of these tools can be introduced and developed through multidisciplinary collaborations and integrated learning opportunities.

2.2 Problem Definition and Motivation

At the heart of creative individuals and collaborations is the ability to define problems, an often childlike curiosity that instills excitement about that which is unknown, and the need to pursue solutions. Individuals and teams must find, discover, or define a problem or problems and develop measurable goals. For example, it is often more fruitful to define the sub-problems than it is to focus on solutions.

2.3 Team Tension

A balance must be sought between a tolerance of diversity and the generative interplay of tension that often leads to conditions in which creativity may flourish. Diversified teams create tension and conflict that can act as catalysts to creative solutions. There is value in nurturing healthy opposition within multi-disciplinary collaborations.

2.4 Multi-disciplinarity and Adhocracy

Typically, individuals require a certain expertise and a certain flexibility for creativity to flourish. This is mirrored within teams where often the most successful are multi-disciplinary and are comprised of experienced and novice members alike; what is lost to rigidity is gained by effective knowledge and mastery of discipline, and what is lost through inexperience is gained in flexibility. The most effective collaborations are non-hierarchical. Granted, non-hierarchical collaborations may be difficult to achieve when combining experienced and novice members, but it is paramount to the process that intellectual equality be respected. Teams should be characterized by an informality and a freedom to say what is on one's mind. Needed are collaborations that foster the interaction of individuals who bring a wide variety of skills and backgrounds to a problem.

Tomorrow ... many solutions will arise when some creative person links existing knowledge in new ways. This will require our separate disciplines to mix more. What may this involve? Certainly, artist must speak with scientist, scientist with laborer and poet, and business people with all. But maybe it's time for scientists to give full value and respect to a new perspective — from the reductionism of the scientific method to the integrative perspective of the social sciences and the humanities.

Dr. Peter Hackett
Edmonton Symposium

The fluidity of organizations must be such that it allows for the rapid formation and dissolution of teams excited by a particular problem. Finally, underlined should be the crucial fact that it is the problem and not the individual that is of central importance.

Examined to this point have been the various implications of viewing creativity and innovation as a process and valuing multi-disciplinarity, both individually and collaboratively. It should be noted that these findings apply equally to individuals, schools, governments, and Granting Councils. In section 3.0, additional general and specific recommendations for fostering environments conducive to the creative process will be presented.

3.0 Recommendations for Governments, Schools, and Granting Councils

3.1 Encourage Multi-disciplinarity

Great benefits and rewards await the nation whose schools and Granting Councils encourage multi-disciplinary study and collaboration, and, hence, creativity and innovation.

Fostering multi-disciplinarity may be accomplished in several ways. Schools should continue to encourage an integrated curriculum with assignments and classroom activities that are problem- rather than discipline-based. Teachers should encourage empathy and explain using real-life examples and analogies from a variety of perspectives and disciplines. In fact, Canadian schools should seek to encourage each of the thinking tools reviewed in section 2.1. These are the creative skills that are needed in artistic, business, university, and research settings.

Encouraging these skills should be continued by universities not only by permitting but also by promoting cross-disciplinary study — for example, by allowing students to combine seemingly unrelated fields such as music and engineering or computer science, history and education; imagination is the only limit to novel and potentially productive combinations. A certain percentage of university appointments should be reserved for multi-disciplinary positions. As well, alternative career promotion protocols should be investigated, since the hierarchical "principle investigator" model can be inappropriate for assessing these types of creative collaborations.

The Federal Government should make additional funds available to each of the Granting Councils (the Canada Council for the Arts, the Canadian Institutes of Health Research, the Natural Science and Engineering Research Council, and the Social Sciences & Humanities Research Council) that are specifically earmarked for truly cross-disciplinary initiatives of the sort likely to encourage creativity and innovation. Such funds would cultivate the networking of researchers from divergent disciplinary backgrounds, not just teams from closely allied disciplines. These funds should be allocated for joint Council sponsorship of projects that span funding bodies. Joint sponsorship would help alleviate the catch-22 scenario that arises when multi-disciplinary projects are not funded because they do not clearly fall under the purview of one funding body or another. For example, a recent research project that sought to examine drama as a way of knowing did not receive funding from the Granting Councils because it was deemed too artistic for SSHRC and too scientific for the Canada Council for the Arts. Such projects are bound to become more frequent as researchers increasingly participate in various multi-disciplinary collaborations and must be addressed.

3.2 Risk, Play, and Great Expectations

Play is the learning by making risky attempts in a set of circumstances where failure is likely, but the consequences are nil.

Dr. Tom Brzustowski
Edmonton Symposium

Creative pursuits require persistence, aiming high, keeping a sense of humour and a sense of play — a childlike curiosity and wonderment is paramount for seasoned scientists and young children alike. Equally important is the need for emotionally and intellectually safe environments in which failure is seen as a successful learning opportunity. Scientists and school children should be encouraged to ask questions and to seek their answers in unlikely directions.

The creative world is the vast abyss beyond what is known. It is the land of total possibility, a blank sheet of paper that is the land of the eagle, where our true power as human beings "dwells," is found, and is expressed.... Einstein, developing his theory of relativity, made this leap of recreating a whole new way of looking at the universe, although the world around him lived in a universe created by Newton. All of these creative people that have created the vast knowledge we have today took the personal responsibility of expressing this marvelous gift that each one of us has. The reason most of us do not exercise or use this powerful gift is that we operate from fear. We are terrified of looking bad, of failing, of being ostracized from the group. Fear is unlimited. It is our fear that keeps us small.

Mr. Douglas J. Cardinal
Edmonton Symposium

Governments, schools, and Granting Councils must recognize the importance of a creatively risk-free environment and the vital role that play plays in fostering arenas and contexts in which learning, creativity, and innovation may flourish.

With this in mind, Granting Councils should allow for numerous small research grants to be allocated to projects that are seen as high-risk, highly likely to fail, but nonetheless low-risk in terms of monetary investment and potentially high-gain in terms of creativity and innovation. Funding should be made available for speculative, seed research that is perhaps not perceived as strategic or valuable within current paradigms. Granting Councils should fund and be more tolerant of high-risk research ventures.

Our schools are taking far too many bright young children, avid learners and naturally creative, and making them dull-eyed, alienated, bullied or bullying, and bored... So, much of the time, does another of our cherished institutions: peer review. And for much the same reason. Chopping off the right-hand part of creativity's bell curve — those rarefied regions of excellence — is seen as the price to pay for the left-hand, underachieving part.

Dr. Peter Hackett
Closing remarks, Edmonton Symposium

3.3 Time to Explore and a Tolerance of Ambiguity

Because creative breakthroughs are often long in the making and require much tolerance for ambiguity, creative research is a long-term investment. Basic research is the well-spring of many innovative products and market applications; however, set directions are frequently unclear, and often require years of exploration before discoveries are made. Creative researchers must be prepared to tolerate, even relish, the long-term and ambiguous nature of their enterprise. Granting Councils must also be willing to tolerate basic research that flows in unspecified directions, is flexible in nature, and that follows ad hoc leads into unexpected areas.

3.4 Individual and Organizational Re-invention

Creative and innovative people are often interested in many and varied areas, frequently working on several differing projects simultaneously and/or changing fields regularly. They are continually challenging and pushing themselves in order to seek answers.

Canadians and Canadian organizations must be prepared to re-examine and re-invent themselves in order to push the boundaries of creativity and innovation. The National Research Council is a good example of an organization that has managed to remain fluid enough to allow for the rapid formation and dissolution of teams that find themselves excited by particular problems.

Universities should also be flexible, perhaps allowing a certain percentage (5%) of any curriculum program to be comprised of ad hoc courses that bypass the standard assessment procedures typically involved when attempting to introduce new courses. Programs should be reevaluated and possibly redesigned every five years; universities should be more willing to embrace new courses and new models. For instance, varying the standard thirteen-week term to create alternative delivery methods and lengths would be a welcomed possibility. Taking a page from the high-technology firms that have recognized the vital importance of seeking and rewarding creative individuals, universities, and Granting Councils should also reward creativity and innovation. Creativity and innovation should be made an overt goal to strive for. "How can we be more innovative?" and "How can we encourage more creativity?" should be questions often asked by administrators, researchers, and educators.

3.5 Creating a Culture of Creativity and Innovation

Recent research into the effects of diversity and multi-culturalism on creativity suggests that multi-cultural policies are headed in the right direction.

> *The diversification of knowledge in its domains surely adds to information. On the other hand, cultural and populational diversity in schools bring additional information.... The more the public system of knowledge is rich and diversified, the more it provides the child with generalisations to confront with her or his particularities.... Education, if conceived as a critical and dialectical process (Popper, 1994) between the individuality of the child and a cumulative and diversified knowledge, is the best means to foster creativity in a society.*
>
> **Dr. Sonia Mansour-Robaey**
> Edmonton Symposium, 2000

As young children encounter diverse environments, the structure and combination of neuronal groups within their young minds are actually physically altered — rendered more likely to produce novel thought patterns and combinations.

These insights and findings coupled with the previously reviewed creative effects of group diversification (see 2.3) make a compelling case for arguing that a creative and innovative culture is a diverse culture.

However, the suggestions to this point are not sufficient. If the Federal Government wants to create a culture of creativity and innovation, then it must also encourage and make additional funds available for exchanges between researchers from diverse cultures. As well, Granting Councils should be rewarding researchers who have been successfully creative and innovative in the past, even if they are now looking to challenge their creative minds by attempting to work in new fields.

Governments should sponsor further colloquia, debates, and public lectures on the topic of creativity and innovation in the arts and sciences. University administrations, business and industry leaders, educators, and the general public need to be made urgently aware of the increasingly crucial importance that creativity and innovation play in the knowledge-based economies of the 21st century. A greater awareness and respect for the cross-pollinary effects and seminal importance of Science–Art interactions must also be generated, for, as observed earlier, science and technology have never flourished in the absence of a similar flourishing of the arts.

4.0 Conclusion

If the Federal Government truly wishes to foster a creative and innovative society in Canada then it must continue to move in the directions outlined in this Report.

> *[In order to create] a society in which people with creativity, vision, and determination — our*

researchers, educators, inventors, and artists — are encouraged to succeed. To succeed here, not somewhere else. To succeed ... in a country and society where new ideas are welcomed and supported.

Speech by the Honourable John Manley
Dec. 1, 1999, to Innovation Canada:
Alliances for the New Millennium

Professor Bentley's call for the formulation of a Task Force on Creativity, Inventiveness, Innovation, and Culture, falling under the combined aegis of the Granting Councils (Canada Council, CIHR, NSERC, and SSHRC), the Humanities and Social Sciences Federation of Canada (HSSFC), the National Research Council (NRC), as well as other interested agencies and organizations, should be acted upon in the near future.

It is especially important that this be done soon because, in the area of creativity and innovation, Canadian society already lags behind Britain, which established a National Advisory Committee on Creative and Cultural Education in 1998, and Europe, where the Creative Europe project has been in existence since 1999 and other projects on the topic of creativity and culture since 1993.

In conclusion, this Report recommends that for Canada to most effectively reap the rewards generated from fostering and promoting a culture of creativity and innovation, an advisory committee should be established to seek additional input from, and provide advice to, a wide variety of stakeholders in the arts, education, sciences, social sciences, medicine, business, and industry from across Canada. The initial task of the Committee should be to explore precisely how the recommendations of this Report may be most effectively implemented and integrated into current organizational, educational, and funding models. Above all, the Federal Government must recognize the importance of creativity and innovation to Canada's future well-being and use all the ways and means at its disposal to ensure a bright future for all Canadians by fostering their powers of creation and innovation.

References

All our futures: Creativity, culture and education (1999). National Advisory Committee on Creative and Cultural Education.

Closing the Gap: Investing in Knowledge for a Better Canada (1998).

creating an innovation culture: Key Challenges and Opportunities as Ontario Moves Ahead in the New Millennium (1998).

Ghiselin, B. (1963). Ultimate criteria for two levels of creativity. In C.W. Taylor & F. Barron (Eds.), Scientific creativity: Its recognition and development (pp. 30–43). New York: Wiley.

Hayes, J.R. (1989). Cognitive processes in creativity. In J.A. Glover, R.R. Ronning, & C.R. Reynolds (Eds.), *Handbook of creativity* (pp. 135–145). New York: Plenum Press.

Hi-tech CEOs Say Value of Liberal Arts is Increasing, April 8, 2000, web version.

Jackson, A.Y. (1976). *A Painter's Country.* Toronto: Clarke, Irwin & Company.

Kuhn, T. (1996). *The Structure of Scientific Revolutions* (3rd Ed). Chicago: University of Chicago Press.

Root-Bernstein, R. (1997). For the sake of science, the arts deserve support. *The Chronicle of Higher Education*, XLIII.

Root-Bernstein, R. (2000). The nature of creativity (Alternate speech for Edmonton Symposium). Personal Communique.

Root-Bernstein, R., & Root-Bernstein, M. (1999). *Sparks of Genius: The Thirteen Thinking Tools of the World's Most Creative People.* New York: Houghton Mifflin.

Speech by the Honourable John Manley, Dec. 1, 1999, To Innovation Canada: Alliances for the New Millennium.

Sustaining Canada as an Innovative Society: An Action Agenda (1997). Ottawa: The Canadian Consortium for Research.

Appendix II. Creativity Conferences, Comments, and Quotes

The Government of Canada's Commitment to Creativity and Innovation

One of the major supporters of the public communications activities associated with the Millennium Conferences series is Industry Canada, a department of the federal government. The department has an important role in the development of policies and programs that affect many national agencies and services in both science, technology, social sciences, and humanities.

> *"Innovation and Creativity are at the forefront of the government's agenda and of the vision of Industry Canada.*
>
> *The foundation of Canada's future prosperity will be new ideas, coupled with the inspiration, creativity, and courage to see them through to the benefit of Canadians."*
>
> **V. Peter Harder**
> *Deputy Minister*
> *Industry Canada*
> *July 2000*

The Future Internet and Collaborative Learning over "Fat Pipes"

Canadian researchers hope to make special contribution to the development of tomorrow's Internet under the long-term collaboration agreement signed between the National Research Council of Canada (NRC) and the National Arts Centre of Canada (NAC) on June 20, 2000, the eve of the Creativity 2000 conference.

An improved Internet based on new "broadband" technology is now being developed by organizations around the world, including the NRC, the Communications Research Centre Canada (CRC), and CANARIE (Canada's Advanced Internet Development Organization). Basically, broadband means that the physical network connections will be carrying more information, and transporting it faster than the Internet of the 20th century. These "fat pipes" will consist of optical fibre cables or, alternatively, radio transmission between fixed antenna and satellites.

> *"A few years ago, it was impossible to imagine today's Internet. Similarly, we do not know what tomorrow's broadband Internet will be like. But we do know that we want it to help people learn together by being able to see and talk to each other, across Canada and the world. In research language, we are exploring tools for 'video-mediated communication' to facilitate 'collaborative learning.' In plain language, this means is that we are using broadband Internet's fat pipes to connect together groups of people at several sites by live video, allowing them to share and experience each others' knowledge and ideas.*
>
> **Dr. Martin Brooks**
> *Group Leader, Interactive Information*
> *Institute for Information Technology*
> *June 2000*

By working with NAC Orchestra Music Director, Pinchas Zukerman, and others with extensive experience with live video teaching, NRC scientists and their colleagues at CRC and CANARIE hope to use broadband learning technologies to build passion for the arts and new tools for education. At Creativity 2000, Maestro Zukerman demonstrated the power of new telecommunications technologies by staging a violin lesson over a satellite link with a student in a remote location whose image was projected on a big screen in the NAC Theatre. Maestro Zukerman not only spoke about the power that technology can play in bringing people together, but also the similarity between artistic endeavours and scientific research.

> *"Music has many, many elements, like the sciences. Everybody divides everything all the time, but it's totally untrue. We are mathematics as they are in science, and they should be the way we are in science."*
>
> **Pinchas Zukerman**
> *Music Director*
> *National Arts Centre of Canada Orchestra*
> *Creativity 2000*

Canada's Role in *The Ascent of Man*

Jacob Bronowski was a mathematician and a man of letters who not only achieved high honours in scientific endeavours but also received critical acclaim for his poetry and prose.

A military reseacher, Bronowski changed careers after seeing the ruins of Nagasaki during a post-war scientific mission to Japan. He pursued the life sciences, the study of human nature, and the evolution of culture. *The Ascent of Man,* the BBC television series that linked science, art, and philosophy in the story of human history, was his last major project. Canada has an enduring association with both the art and technology of this epic TV series through its computer animation sequences produced by the then world-leading centre for such work at the National Research Council (NRC) in Ottawa. Coincidentally, Canada's pioneering work in computer animation was often held up in the Millennium Conferences as an example of the creative power of a merger of arts and sciences.

> *"In every age there is a turning point — a new way of seeing and asserting the coherence of the world."*
>
> **Jacob Bronowski**
> *The Ascent of Man, 1973*

Merge Arts and High Tech with a Career in Industrial Design

One occupation that provides artistic and visually creative young people with an opportunity to participate in the high-technology, advanced manufacturing, and engineering worlds is industrial design.

The study of industrial design typically involves art, science, engineering, business, ergonomics, social issues, environmental studies, and psychology. The School of Industrial Design at Carleton University, a supporter of the Millennium Conferences on Creativity in the Arts and Sciences and one of the leading schools in North America offering undergraduate programs in Industrial Design, has, since 1995, given secondary school students across Canada an opportunity to experience the field through its annual High School Industrial Design Competition.

> *"I much admire the Creativity initiative. Long may it continue."*
>
> **Professor Brian Burns**
> *Director, School of Industrial Design*
> *Carleton University*
> *September 2000*

The Professor of Literature Who Changed High Tech

American Geoffrey Moore holds a bachelor's degree from Stanford University and a Ph.D. from the University of Washington, both in literature, and worked as an English professor for many years before turning his mind to the high-technology industry. His books have not only been successful as best sellers, they have articulated visual metaphors that facilitated the development of the high-technology industry through their wide acceptance as apt descriptions of the challenges facing the industry and strategies for confronting them. Today, he is a consultant, leading authority on high-tech growth, and venture capitalist, who came to Ottawa in September 2000 to announce that his firm was investing $6.8 million in Trillium Photonics, a firm created by National Research Council (NRC) researchers.

During his visit to Canada, he spoke as part of the third Asia–Pacific Economic Co-operation (APEC) R&D Leaders Forum, organized, in part, by the Ottawa Regional Innovation Office. Commenting on the imminent election of a one-tier regional government in the Ottawa region, Moore urged a co-ordinated, balanced approach to high-tech economic development reflecting upon his blended background in technology and arts.

> *"they should immediately go to San Jose to see how not to do it."*
>
> **Author Geoffrey Moore**
> *Ottawa*
> *September 2000*

The Artistic Side of Einstein's Brain

In 1999, Canadian scientists announced findings that may help explain Albert Einstein's creative genius. His brain, which had been removed and preserved after his death in 1955, was compared by Dr. Sandra F. Witelson and her colleagues at McMaster University in Hamilton in the shape and size to the brains of dozens of men and women of average intelligence. In general, Einstein's brain was the same as all others except in one particular area — the region said to be responsible for both mathematical thought and the ability to think in terms of space and movement was more developed. Einstein was known for describing his scientific thinking in terms of images of a visual kind.

> *"We never solve our problems if we stay at the same level we were when we created them."*
>
> **Albert Einstein**

C.P. Snow Wasn't a "Bad Guy"

Many participants in the Millennium Conferences on Creativity in the Arts and Sciences made reference to the works of Charles Percy Snow, the British novelist and scientist whose works included "The Two Cultures and the Scientific Revolution" given as a Rede Lecture at Cambridge in 1959 and published as a small book. In it, Snow (1905–1980) highlighted the chasm between literature and science from the perspective of someone who knew both worlds.

> *"As an event to promote the coming together of the arts and sciences, Creativity 2000 was excellent. I'd give it an A+ — both for the intention and execution. The choice of speakers (both for their usually justifiably high profiles, prestige, and their skill as speakers) was first rate. But I would highly recommend to those who enjoyed the event or have an interest in this issue that they read C.P. Snow's essay. Contrary to the impression left by some, Snow wasn't the bad guy; he would have been the first to propose and promote an event like Creativity 2000."*
>
> **Ken Stange**
> *Author and Lecturer on*
> *Creativity and the Psychology of Art*
> *Nipissing University College, North Bay, Ontario*

Arthur C. Clarke and Creativity 2000

> *"Creativity is a form of Play."*
>
> **Arthur C. Clarke**
> *June 21, 2000*

Creativity 2000, the centrepiece conference on creativity in the arts and sciences staged at the National Arts Centre (NAC) on June 21st, was memorable for many reasons, including something that *did not happen*.

Author and futurist Sir Arthur C. Clarke did not have a chance to share his thoughts on the creative process as he had hoped.

Clarke, whose live participation via satellite was arranged by the Canadian offices of the British Council, was cut short in his interview with

Canadian actor and film maker Don McKeller when satellite time ran out.

Before the link to Sri Lanka, where Clarke has lived for four decades, ended, the audience and webcast viewers of Creativity 2000 were able to see Clarke on the big screen in the NAC Theatre, laughing and joking with his interviewer about his Wizard of Oz-style presence and the image of "Big Brother being watched." He told the audience, in fact, that

> *"Arthur C. Clarke is quite busy, and this is one of his clones."*

His many fans in the audience were pleased to see him in good health and cheerful. Clarke, now well into his 80s, was at the time often dependent on canes or a wheelchair. In the comments he did make on the subject of the conference, Clarke echoed the view that creativity is "play" which draws upon our childlike qualities as well as intellect, knowledge, and skill.

"*Creativity is a form of Play*" Clarke said to the Canadian audience, suggesting that in the arts "*where anything goes*" it may be more apparent, but that "*science is also a form of play: play restrained by the physical universe.*" In the weeks following Creativity 2000, organizers of the conference had a number of e-mail exchanges with the writer to apologize for the interruption in the interview and to thank him for his participation. Clarke responded to all messages despite his busy schedule.

At the time, he was not only involved in writing for books and magazines, but also addressing the numerous interview requests that arrived daily via e-mail and fax to his Sri Lankan home. He told Creativity 2000 organizers, for example, that he was overwhelmed by media and science leaders seeking his opinion on the year 2000 discovery of evidence of water on Mars.

He asked for a copy of the video of Creativity 2000 and suggested that these of his famous quotes might be relevant to this Report and the creativity discussions that he missed.

> *"We have to abandon the idea that schooling is something restricted to youth. How can it be, in a world where half the things a man knows at 20 are no longer true at 40 — and half the things he knows at 40 hadn't been discovered when he was 20?"*

> *"I don't pretend we have all the answers. But the questions are certainly worth thinking about."*

> *"The only way of finding the limits of the possible is by going beyond them into the impossible."*

> *"If an elderly but distinguished scientist says that something is possible he is almost certainly right, but if he says that it is impossible he is very probably wrong."*

In his last e-mail on the subject, Sir Arthur added, however, that of all of his thoughts relevant to the Creativity 2000 discussions,

> *"...the most famous, and I think most important: Any sufficiently advanced technology is indistinguishable from magic..."*

Clarke, considered the Godfather of Science Fiction writers, has authored over 80 books, but he recommended "the latest edition of (his book) *Profiles of the Future*" as the one that would have information of most interest to the participants of Creativity 2000.

Britain Helps Link Arts and Sciences Around the World

Creativity through the links between arts, sciences, and education is well recognized in Britain. Not surprisingly, one of the early international supporters of the Millennium Conferences on Creativity in the Arts and Sciences was the British Council, an independent international agency that encourages cultural, scientific, technological, and educational co-operation between Britain and other countries.

The Council, which has been active in Canada since 1959, is co-located with the British High Commission in Ottawa and the British Consulate-General in Montréal. Its programs include a collaboration with the National Research Council to jointly support public sector collaborative research, as well as student and staff exchanges in science and engineering.

> *"We were very impressed by the discussions at Creativity 2000 and proud to be associated with it."*

Keith F. Preston, *Ph.D.*
Fund and Secretariat Administrator
NRC–BC Fund
The British Council Canada

An Asian Perspective on the Need to Link Arts and Sciences

The National Science Council (Taiwan) was one of the international supporters of the Millennium Conferences on Creativity in the Arts and Sciences and a sponsor of the *Creativity and the World Banquet* on June 20, 2000. The NSC has a long-standing commitment to building bridges between the arts and sciences, as its mandate not only includes support for scientific and engineering research and the administration of science-based industrial parks, but also support for research in journalism, philosophy, literature, and other fields of humanities and social sciences. This broad perspective is reflected in the following comments.

> *"Few artists really understand science. Likewise, many scientists do not know much about art either. The question is — Is there any connection between art and science? One clue I found is 'algorithm,' which exists in both art and science. While in arts algorithms are the hidden rules*

behind many works of art, they are procedures for solving problems or backbones for writing computer programs in science or engineering. One good example is a methodology adopted by many researchers to analyze the generation of images and the composition of famous paintings in order to reveal the secret in various styles of art creation.

Technologies that merge art and science, such as hypermedia and digital libraries/museums, are being used to preserve our precious artistic knowledge and cultural legacy. For example, the National Science Council (NSC) Taiwan is sponsoring a 'digital museum' project as part of the 'Greeting a New Millennium — A 21st Century Science Development Program with Concern for the Humanities as a Main Theme.' This project establishes a model website which weaves together culture, art, science, and technology and promotes the research in related technologies for Digital Library System, such as metadata, watermark and multimedia technologies.

In Taiwan, more researchers in humanities have applied technologies such as 3D and virtual reality in their research. NSC Taiwan has recently sponsored a researcher in the Department of Chinese Literature at National Taiwan University who used 3D simulations to recreate a video show of an ancient Chinese wedding ceremony for which only the written records can be found today.

Artists in the 21st Century also can no longer evade the powerful impacts from information technologies, such as digital images, 3D animations, virtual reality, and interactive designs. The development and application of various algorithms also declares this emerge of digital arts. In November 2000, an exhibition 'Art Future 2000' held in Taipei was a big festival for both artists and scientists. Famous local and international artists were invited to display their works in the exhibition, and to participate in the Digital Art Contest. International artists and experts in hi-tech also attended a symposium in Taipei to discover the new possibilities in future arts and to discuss the future integration of art and science for improving the spirit in human life. We always welcome elite artists and scientists around the world to come to Taiwan and share with us their new thoughts and precious creative experiences."

Dr. Cheng I Weng
Chair
National Science Council (Taiwan)
December 2000

A Science for Artists and Philosophers

For those interested in learning more about science and technology, the field of astronomy offers many opportunities and issues that seem to cross boundaries and appeal to artists, humanists, young children, and the general public as well as scientists from many disciplines.

> *"The health of a culture and economy in the information age depends on its ability to contribute to basic research, to compete technologically with other nations, and to educate its citizens both in science and the arts. The pursuit of astronomy and astrophysics furthers all of these goals.*

Almost one million people annually visit Canadian observatories and planetaria ... At the university level, approximately 10,000 Canadian students elect to take non-specialist courses in astronomy each year; for most of them, it will be the only formal university contact they will have with science."

The Origins of Structure in the Universe: Canadian Astronomy and Astrophysics in the 21st Century
June 2000

Educator's Three Messages of Caution to Innovation Leaders

In December 1999, the **Canada Foundation for Innovation** and its partners staged the Innovation Canada Conference in Ottawa, which attracted research and industrial leaders from across Canada for a gathering that explored ways to promote economic growth and innovation through new alliances.

In one of the sessions, NAC Director and CEO Peter Herrndorf spoke on the power of the arts in a high-tech world saying, *"Art stimulates the brain's synapses; they get us thinking about things in different ways. The artistic capacity of a nation will have a direct impact on its ability to compete in a knowledge-based economy."*

He was followed by the then-Dean of Education at Queen's University who warned the audience of industry and research leaders on three points.

"Let me begin with three messages of caution in this midst of this celebration of creation and innovation.

Peter Herrndorf talked about the importance of bringing together the scientific, technological, literary, and artistic disciplines as equal partners in innovation. I agree with his view, and in doing so, I would like to reinforce the importance avoiding distinctions of status as well — the valuing of one enterprise over another.

- *We must not value science more than art nor art more than science — no more than we would give up one half of the brain for the other. The same goes for valuing research over teaching or the private sector over the public or vice versa. What we must value instead are the differences between the arts and sciences, the humanities and the social sciences, as well as the natural alliances between them. To do any less would enforce what Social Sciences and Humanities Research Council President Marc Renaud calls a 'culture of poverty.' And poverty in any group impoverishes us all.*

- *My second concern is that we not overlook the intangible measures of success. In addition to the performance variables and deliverables and the economic performance indicators that (Royal Bank CEO) Mr. Cleghorn described so convincingly, there must be room for acknowledging the joy that is the internal sign of creativity, and the beauty that is the sign of creativity for all to see, whether that beauty manifests itself as a sculpture or in the elegance of a mathematical proof or — as some of us saw this morning — in the complexity of the gear on the forehead of an ant as revealed by synchrotron light. All of these things are beautiful in their own right.*

- *The third caution is this. We must not interpret lack of progress and supreme frustration as signs of failure. This is risky work and it is hard work, particularly in the early days. It will be hard, for example, to make the Canadian Institutes for Health Research work. The work will be hard, not because of a lack of natural alliances between the social sciences and biomedical disciplines, or between the government, universities, and private sector, for that matter, but because there is a need to break down the silos while at the same time, honouring specializations as a new research culture is formed. As we move forward with CIHR and other initiatives like it, it will be imperative to remind ourselves that natural processes are hard work — a natural process, and if I remember correctly, birth is very hard work indeed.*

Speaking of birth, it is no coincidence that — quite independently — most of tonight's speakers have asked us to cast our minds back to Renaissance.

La Renaissance a été un époque extraordinaire en matière de réalisations et de célébration du génie et du potentiel humains. Synonymous with the Renaissance is the figure of Leonardo da Vinci. Five hundred years from now, people will still be talking about him. But perhaps not everyone knows that Leonardo was granted a position in Milan in his capacity as a musician, although the services he also offered included those of architect, painter, sculptor, and engineer. Indeed, Leonardo was permitted to do his science and engineering so long as it did not interfere with his art. One wonders if he were living now, if he would be permitted to do his art so long as it did not interfere with his science — an equally bizarre restriction, for it is each that feeds the other.

Let me now turn to a contemporary Renaissance figure, the Canadian astronaut Julie Payette, who is a scientist, engineer, linguist, scuba diver, jet pilot, triathlete, soprano singer, and pianist. In a conversation I had with Dr. Payette when she was recently recognized at Queen's with an honorary doctorate, I asked her about her many pursuits. For her, they are inseparable. She said to me — with joy and a twinkle in her eye — 'Everything I have done has led to this.'

I believe that we are ripe for a new Renaissance, a new time of unprecedented achievements, of countless Julie Payettes and Pinchus Zuckermans, of celebration of human spirit and potential, of education, of hard work, of beauty, and of joy. A time to value the arts, the social sciences, the humanities, the sciences, and to nurture the natural alliances between them.

This is creation. This is innovation."

Professor Rena Upitis
(then) Dean of Education
Queen's University
Innovation Canada
December 1999

Teach Science Fiction as Both Good Art and Good Science

As professionals, both scientists and writers of science fiction extrapolate on the current state of science and technology, albeit for different reasons.

There are historical examples of fiction writers who have made reasoned and accurate predictions of the advent of certain technologies, often well in advance of their scientific counterparts (Jules Verne and the space race, H.G. Wells and the atom bomb, Arthur C. Clarke and geosynchronous satellites). There also exist significant counterexamples, such as the Internet, which were largely unpredicted by both communities.

There is a mutual reliance between the two disciplines. Today, in an effort toward enhanced realism, which is in part motivated by the increase in scientific literacy of their audience, fiction writers often draw inspiration from the scientific literature, sometimes interfacing directly with the scientists themselves. For their own part, scientists tend to be avid (if sometimes closet) science fiction fans. There are also those people who bridge the disciplines, working by day in scientific and technical professions while also authoring fiction at night.

The Canadian Science/Science Fiction Conference in September 2000 investigated the broad relationships between science and science fiction, including the degree to which science fiction anticipates science. The Conference attracted high-profile writers such as the "Dean of Canadian Science Fiction" Robert J. Sawyer and science journalists such as CBC's Bob McDonald. One of the participants, Julie E. Czerneda, described not only how she brings her fascination with biology to her science fiction works, but also to the classroom. Her books often target young readers with original, illustrated SF stories based on elementary science curriculum. The companion resources have experiments, integration ideas, background information, and other tools to help teachers use the speculation and creativity of science fiction literature to encourage students to critically assess the scientific information in it.

"Art imitates life ... or is it the other way around?"

Dr. Michael Greenspan
Organizer
Canadian Science/Science Fiction
Conference 2000

Astronaut Tells Students How to Travel Through Time

On June 14, 1999, former Canadian astronaut Major Mike McKay addressed over 200 students from local schools in an auditorium of the historic National Research Laboratories Building at

100 Sussex Drive, Ottawa. Major McKay made a presentation about the science of space and time and encouraged the students to use their imaginations to travel through space and time and to make their dreams come true. His presentation was combined with the unveiling of the Canadian Millennium Clock, a clock linked directly to the national cesium beam atomic clocks and Canada's official time. Canadian scientists built the world's first continuously operating primary atomic clock in the 1970s. The Canadian Millennium Clock celebrates this achievement.

> *"The best way to travel through time is to use your imagination."*
>
> **Major Mike McKay**
> *Addressing students*
> *June 1999*

How Canadian Technology Helps Renaissance Art

In October 1999, experts in 3D imaging and modeling technologies from around the world met in Ottawa for a series of events hosted by scientists at the National Research Council of Canada (NRC). NRC's Institute for Information Technology (IIT) in Ottawa is the recognized world leader in a three-dimensional imaging technology that has seen applications in fields ranging from art preservation and computer animation to engineering design and medicine. Canadian high school students have worked with the researchers to reconstruct 3D computer models of buildings and cities in projects that blend the teaching of leading edge computer skills with historical research in the 3D Historical Cities Project.

The October 1999 3D technology events were kicked off with the Canada–Italy Workshop on Heritage Applications of Digital 3D Imaging. This workshop highlighted the outstanding benefits of high-resolution three-dimensional imaging of important architectural structures, archaeological sites, sculptures, and museum collections. NRC scientists have participated in a number of many international programs including the international Digital Michelangelo Project to gather detailed measurements of Michelangelo's sculptures, including the celebrated David, and a collaboration with the University of Padua, Italy, to demonstrate a portable 3D camera in a museum setting. The use of 3D technologies in art and heritage applications were among the success stories of scientific partnerships with the artistic community highlighted in the media in connection with the Millennium Conferences.

"We decided that we would like a higher-quality gold standard for our scan. The best laser scanner in the world is at (Canada's) National Research Council."

Professor Marc Levoy
Stanford University
Head, Digital Michelangelo Project
October 1999

A Conference for Dancers at a Research Laboratory

In mid-November 1999, the National Arts Centre of Canada and the National Research Council in association with the School of Dance and the Canadian Medical Association sponsored three satellite downlink sessions of "Not Just Any Body — a Global Conference to Advance Health, Well-Being, and Excellence in Dance and Dancers." Members of the Ottawa area dance, arts, and health community attended the sessions in the auditorium of an NRC laboratory building. The main presentations in Toronto and the Hague were viewed via satellite on a large screen, and the participants then discussed the links between creativity and health and the challenges of merging artistic and technical excellence in dance.

The speakers in Toronto included creativity expert Mihaly Csikszentmihalyi, the author of *Flow: The Psychology of Optimal Experience*. He and others spoke on how individuals and organizations can achieve peak performance. Csikszentmihalyi described his theory that people experience "flow" when immersed in creative work that they love. This concept of flow refers to a state of happiness that an individual experiences when he is engaged in a well-defined task that is both challenging and within his capabilities.

"Dancers must marry art and technique to be creative."

Audience participant in
"Not Just Any Body" downlink session

Buffy Sainte-Marie — an Artist Who Promotes Science

The National Aboriginal Career Symposium seeks to encourage students to stay in school, focus on career-oriented educational choices, and maintain their cultural identity within the workplace. In October 1999, the fourth such event was held in Ottawa.

Native American astronaut Lt. Commander John Herrington, NASA, and Academy Award-winning songwriter, singer, visual artist, and educator Buffy Sainte-Marie delivered keynote addresses at the opening ceremonies. Buffy Sainte-Marie also hosted a workshop in connection with the

Cradleboard Teaching Project, a program to develop unique education partnerships as well as curriculum materials such as the CD-ROM "Science: Through Native American Eyes."

A teacher before she ever started singing, Sainte-Marie has continually used her talents in art, music, and cutting edge technology to educate both on stage and in the classroom. Buffy Sainte-Marie's work exemplifies the harmony and creativity that flows between the arts and sciences. Her participation in the Symposium was sponsored by the Millennium Conferences on Creativity in the Arts and Sciences.

> *"to bring out the best of our hearts into our presence ... this Navajo Elder (said), just remember, computers, telephones, fax machines, whatever people may invent ... nothing takes the place of the fire ..."*
>
> **Buffy Sainte-Marie**
> *National Aboriginal Career Symposium*
> *Ottawa*
> *October 1999*

Creativity and the Economics of the Future

On Friday, October 15, 1999, the NRC Entrepreneurship Program in collaboration with the Millennium Conferences on Creativity in the Arts and Sciences staged a special presentation on the economics of the future and the role of creativity.

As part of this special Lunch-time Speaker Series event, Dr. Sherry Cooper, Senior Vice-President and Chief Economist of Nesbitt Burns, discussed creativity and innovation in a talk entitled "Profit and Prosper in the New Millennium."

> *"Creativity is important in the new economy as we need to create more Ottawa-style high-tech regions and to focus on creating new businesses industries rather than propping up those that are fading."*
>
> **Dr. Sherry Cooper**
> *Economist*
> *October 15, 1999*

Rock and Roll — Physics and Math

In mid-February 2000, 150 Ottawa students were given the formula for becoming a Rock Star. The key, according to Paul Hoffert, former researcher and co-founder of the band Lighthouse, is to use your imagination and to mix science and technology with rock 'n roll. Hoffert spoke to an audience of primarily music students, who were eager to learn how to break into the international entertainment business.

Hoffert encouraged these aspiring musicians to be technology savvy, and not to limit their focus to just the arts.

Hoffert spoke of the "Renaissance Man," integrating all skills together and working toward a general knowledge base. Hoffert encouraged students to learn how to capture their sound and image digitally. Hoffert's wisdom was backed up by current Lighthouse singer Dan Clancy. He also stressed the value of learning the technical side of music, which opens up new possibilities such as independent CD production and marketing on the web.

> *"In our society, arts and science have been very separated and essentially all the specialization that's happened isn't working. The digital age that we're coming to is so different from the industrial revolution that we need a different way to think about things."*
>
> **Paul Hoffert**
> *Founder of rock group Lighthouse*
> *Former Chair of Ontario Arts Council and Researcher*

"The Einstein of Dance"

Canadian choreographer Édouard Lock's approach to his work and his impact on the world of dance was compared to Albert Einstein's influence on physics in a newspaper profile marking the June 2000 Canada Dance Festival. The Festival, co-produced by the National Arts Centre, included a tribute to Lock, in honour of the 20th anniversary of his innovative company, La La La Human Steps. A producer of contemporary dance, Jack Udashkin, was quoted as saying that Lock, whose innovative dance troupe has entertained in large stadiums around the world,

> *"... is like (James) Joyce and Einstein in terms of their influence on the forms in which they worked, and the way in which they used older forms to develop new forms."*

Fractals and the Underpainter

Creativity 2000 sought to encourage new interactions between the arts and sciences through discussions that brought exceptional individuals in both worlds together. In a morning session entitled "Infinite Possibilities," author and poet Jane Urquhart shared the stage with a giant screen, which allowed for live interaction with mathematician and internationally recognized Father of Fractal Geometry Benôit Mandelbrot. Urquhart, whose poetry and novels (*The Whirlpool*, *Changing Heaven*, *Away*, and *The Underpainter*) have been published around the world, presented a reading and commented on the theme of the conference from her personal experiences.

> *"Both scientists and artists are rooted in the perceived world ...*
>
> *creativity is important in the sciences and in the arts, and I think that had I been a little more exposed to that as a student, I think that I would have been a much better student, probably a much better English and History student, as well as a Maths and Sciences student."*

Jane Urquhart
Poet and Author
Creativity 2000

The Sterling Professor of Mathematical Sciences at Yale University and IBM Fellow Emeritus, Mandelbrot has received international recognition and honours for his work and influence on geometry, physics, computer science, finance, and art. Yet, he told the audience that his study of fractals, the geometry of patterns at different scales, has had an impact beyond anything he imagined.

> *"Gradually, we had more and more references to fractality, not from scientists (they came later), but from the artists, the musicians, the poets ..."*
>
> *"Very great contemporary musicians came to me to tell me 'Do you realize that your work has helped us to understand music?'*
>
> *And I thought 'How can that be? I never imagined that possible!' They told me of how they thought of a piece of music as being whole with parts, and parts, and parts, the way in which we are taught about it, in a certain sense, without the proper language, was very much in this context."*
>
> *"There is no marker on the earth stating that you are leaving mathematics and entering physics, you are leaving physics and entering finance, into music; there are no markers at all; it is all one big unified cloth."*
>
> *"The people [Arts majors] have an extraordinary thirst of mathematics if it is presented to them not primarily as the plumbing of mathematics, which is essential but which is rather dry and indeed very difficult, but as a way of thinking about patterns and structure which is not specific to thinking about one pattern of pure mathematics, physics, or engineering, but which transcends into everything."*

Benôit Mandelbrot
Father of Fractal Geometry
Creativity 2000

Trying to Sum it All Up

The daunting task of summing up the day's discussions fell on the shoulders of author and philosopher Mark Kingwell (who co-hosted the Creativity 2000 conference with National Research Council Secretary General Lucie Lapointe) in a conversation with Sir John Maddox, the theoretical physicist turned author who is widely credited for the prestige of the international scientific journal *Nature* which he oversaw as Editor for more than 20 years.

Professor Kingwell approached his task in part by noting some of the words, names, and phrases that stood out: *"I couldn't help picking up a few tidbits during the day; things like curves and straight lines, hexagons, violins and music, Venice, Mozart, Plato, Einstein, Buckminster Fuller, Glen Gould, toys and play and children, wonder, the weather, Jane's book, flunking out of school, staying in school, what's wrong with the schools, Star Wars, risk and reward, and finding patterns ... Interesting, or useful, or beautiful, or elegant — these are some of the adjectives we use about both art and science."*

> *"Art and science are both conversations, and they're aspects of a larger human conversation of which this conference is a small part."*
>
> **Mark Kingwell**
> *Author and Philosopher*
> *Creativity 2000*

For his part, Sir John reiterated the need for "courage" and the view that creativity is "doing something for the first time in an imaginative way," adding that it has to be "in a way that commands the respect of the community in which you work." He also echoed the importance of study of the human brain and its relationship to the more elusive "creative mind," saying, "we are a long way away from understanding how thinking is actually accomplished in the head. It will, of course, be a great day when that is understood, but that day is a long way away, in my opinion." He added his hope that the conference would have an impact.

> *"As a visitor, I wish that the conference is aftermathed well."*
>
> **Sir John Maddox**
> *Scientist and Author*
> *Creativity 2000*

Appendix III. Creativity 2000 Programme

NATIONAL ARTS CENTRE
OTTAWA
JUNE 21, 2000

Presented by: The National Research Council of Canada, the National Arts Centre, and the Canada Council for the Arts

CO-HOSTS

- **Mark Kingwell,** Associate Professor of Philosophy, University of Toronto
- **Lucie Lapointe,** Secretary General, National Research Council

8:00

REGISTRATION, FOYER, NATIONAL ARTS CENTRE

Performance by the School of Dance — beginning at 8:40

9:00

OPENING CEREMONIES

Her Excellency the Right Honourable Adrienne Clarkson
The Honourable Herb Gray,
Deputy Prime Minister
Peter Herrndorf, Director General and CEO, National Arts Centre
Dr. Arthur J. Carty, President,
National Research Council

9:20

WIRED MUSINGS

A conversation between Sir Arthur C. Clarke, author and futurist, in Sri Lanka, and Don McKellar, actor, writer, and film maker, via satellite
Sponsored by: The British Council

9:50

INFINITE POSSIBILITIES

Introduction: Dr. Peter Hackett, Vice-President of Research, National Research Council

Participants: Benoît Mandelbrot, "Father" of fractal geometry
Pinchas Zukerman, Music Director, National Arts Centre Orchestra
Jane Urquhart, author and poet

Facilitator: Jean Gagnon, Director of Programs, Daniel Langlois Foundation for the Arts, Science and Technology

Dance Solo: Benoît Lachambre (choreography by Meg Stuart)

Sponsored by: Canada Foundation for Innovation

LUNCH

Performances by the School of Dance and actor Peter Froehlich, and works on loan from the Canada Council Art Bank will be presented throughout the lunch period.

13:00

THE CREATIVE MIND

Introduction: Dr. Shirley L. Thomson, Director, Canada Council for the Arts

Participants: Dr. Albert Aguayo, Director of the Centre for Research in Neuroscience, McGill University
Dr. Margaret Boden, Professor of Philosophy and Psychology, University of Sussex
Catherine Richards, artist

Facilitator: Dr. Mark Kingwell, Associate Professor of Philosophy, University of Toronto

Dance Solo: Tedd Robinson, choreographer and performer

Sponsored by: Canadian Institutes of Health Research

15:20

RECONFIGURATIONS: STRUCTURE AND SPACE

Introduction: Peter Herrndorf, Director General and CEO, National Arts Centre

Participants: Sir Harry Kroto, Nobel laureate
Douglas Cardinal, architect

Facilitator: Don McKellar, actor, writer, film maker

Dance Solo: AnneBruce Falconer (choreography by Louise Bédard)

Sponsored by: Natural Sciences and Engineering Research Council

FUTURE INSIGHTS: "What does it all Mean?"

Participants: Dr. Mark Kingwell, Associate Professor of Philosophy, University of Toronto
Sir John Maddox, Physicist

17:50

Concluding address: Dr. Arthur J. Carty, President, National Research Council

18:00

Reception

Organizing Committee — Creativity 2000

Co-Chairs

Dr. Michel Brochu
Director, Executive Offices Secretariat
National Research Council

Mr. Bernard Geneste
Senior Director, Corporate Services and
Corporate Secretary
National Arts Centre

Members

Ms. Kelly Ann Beaton
Director, Communications
National Arts Centre

Dr. John Bradley
Institute for Research in Construction
National Research Council

Mr. Martin Brooks
Head, Interactive Information Group
Institute for Information Technology
National Research Council

Mr. Richard I. (Dick) Doyle
Project Manager
Millenium Conferences on
Creativity in the Arts and Science
National Research Council

Mr. Keith Kelly
Director, Planning and Research
Canada Council for the Arts

Mr. Robert Laliberté
Director, Communications
National Research Council

Dr. Dennis Salahub
Director General,
Steacie Institute for Molecular Sciences
National Research Council

Ms. Nicole Sarault
Manager, Conference Services
National Research Council

Ms. Katherine Watson
Producer
Creativity 2000

Secretary

Ms. Sylvie Brunette
Business Relations Office
National Research Council

Index — Individuals Quoted/Cited (not listed in Contents)